Whitfield's Electrical Craft Principles, 6th Edition

Other related titles:

You may also like

- PWTP182B | Guide to Temporary Electrical Systems, 2nd Edition: Design, management and deployment | 2025
- PWRP001P | Open Combined Protective and Neutral (PEN) Conductor Detection Devices (OPDDs) – OPDDs for use in electric vehicle charging applications for household and similar installations | 2024
- PWR1820B | Requirements for Electrical Installations, IET Wiring Regulations, Eighteenth Edition | 2022
- PWGO182B | On-Site Guide, 8th Edition | 2022

We also publish a wide range of books on the following topics:
Computing and Networks
Control, Robotics and Sensors
Electrical Regulations
Electromagnetics and Radar
Energy Engineering
Healthcare Technologies
History and Management of Technology
IET Codes and Guidance
Materials, Circuits and Devices
Model Forms
Nanomaterials and Nanotechnologies
Optics, Photonics and Lasers
Production, Design and Manufacturing
Security
Telecommunications
Transportation

All books are available in print via https://shop.theiet.org or as eBooks via our Digital Library https://digital-library.theiet.org.

ELECTROMAGNETIC WAVES 086

Whitfield's Electrical Craft Principles, 6th Edition

Volume 1

Terry Grimwood and John Whitfield

The Institution of Engineering and Technology

About the IET

This book is published by the Institution of Engineering and Technology (The IET).

We inspire, inform and influence the global engineering community to engineer a better world. As a diverse home across engineering and technology, we share knowledge that helps make better sense of the world, to accelerate innovation and solve the global challenges that matter.

The IET is a not-for-profit organisation. The surplus we make from our books is used to support activities and products for the engineering community and promote the positive role of science, engineering and technology in the world. This includes education resources and outreach, scholarships and awards, events and courses, publications, professional development and mentoring, and advocacy to governments.

To discover more about the IET, please visit https://www.theiet.org/.

About IET books

The IET publishes books across many engineering and technology disciplines. Our authors and editors offer fresh perspectives from universities and industry. Within our subject areas, we have several book series steered by editorial boards made up of leading subject experts.

We peer review each book at the proposal stage to ensure the quality and relevance of our publications.

Get involved

If you are interested in becoming an author, editor, series advisor, or peer reviewer please visit https://www.theiet.org/publishing/publishing-with-iet-books/ or contact author_support@theiet.org.

Discovering our electronic content

All of our books are available online via the IET's Digital Library. Our Digital Library is the home of technical documents, eBooks, conference publications, real-life case studies and journal articles. To find out more, please visit https://digital-library.theiet.org.

In collaboration with the United Nations and the International Publishers Association, the IET is a Signatory member of the SDG Publishers Compact. The Compact aims to accelerate progress to achieve the Sustainable Development Goals (SDGs) by 2030. Signatories aspire to develop sustainable practices and act as champions of the SDGs during the Decade of Action (2020–2030), publishing books and journals that will help inform, develop, and inspire action in that direction.

In line with our sustainable goals, our UK printing partner has FSC accreditation, which is reducing our environmental impact to the planet. We use a print-on-demand model to further reduce our carbon footprint.

Published by The Institution of Engineering and Technology, London, United Kingdom

The Institution of Engineering and Technology (the "**Publisher**") is registered as a Charity in England & Wales (no. 211014) and Scotland (no. SC038698).

First published: 1974
Second edition published: 1980
Third edition published: 1989
Fourth edition published: 1995 (0 85296 811 6)
Reprinted: 1997, 1999, 2001, 2005, 2006 and 2007
Fifth edition published: 2008
Sixth edition published: 2025

The Institution of Engineering and Technology
Futures Place,
Kings Way,
Stevenage,
Herts, SG1 2UA
United Kingdom

www.theiet.org

British Library Cataloguing in Publication Data

A catalogue record for this product is available from the British Library

ISBN 978-1-83724-254-2 (Volume 1 hardback)
ISBN 978-1-83724-255-9 (Volume 1 PDF)

ISBN 978-1-83724-256-6 (Volume 2 hardback)
ISBN 978-1-83724-257-3 (Volume 2 PDF)

Typeset in India by MPS Limited

Cover image credit: Multi coloured twisted cables, Jonathan Kitchen/DigitalVision via Getty Images

Contents

John Whitfield Obituary

John Whitfield was born in 1930. As befitting people (even young people) who lived through the war, life was tough. He rose within an electrical contracting company that carried out work throughout England, eventually managing it. By his own admission, however, at the time he was 'burning the candle at both ends'. This coincided with him being diagnosed with multiple sclerosis. Typifying his positive attitude, rather than letting this beat him, he threw himself into technical education where he became a positive role model for many electricians learning their trade. This wide experience led him to be the author of many books. He continued to work on his books and technical papers until his death in 2019 at the age of 89 years, passing away peacefully. He was a beloved husband to Joyce and leaves behind his daughter Lesley and son Michael.

About the authors

Terry Grimwood MA, PGCE, is the co-author of numerous electrical and engineering textbooks as well as five novels and many novellas, plays and short stories. His industry experience ranges from electrical installation through to project management, quality assurance and technical writing. He taught electrical installation and technology for twenty years at both Otley College (Suffolk) and Oaklands College, St Albans (Hertfordshire), before becoming a full-time writer.

John Whitfield spent 30 years as a lecturer at Norwich City College of Further and Higher Education, where he gained very wide experience in teaching electrical installation, electrical craft practice, electrical technician's courses and the Ordinary and Higher National Certificate and Diploma courses. His insight into the needs of students in technical education and his wide experience of teaching methods enabled him to base the earlier editions of this book on well-proven techniques. John continued to work on his books and technical papers until his death in 2019 at the age of 89 years.

Chapter 1

SI units and basic mathematics

1.1 Introduction

Mathematics underpins electrical science, technology, design, and inspection and testing. It is assumed that you will already have a grasp of basic mathematical principles; however, to ensure a full understanding of the subjects covered by both Volumes 1 and 2, this chapter provides revision and a reference source for the calculations you will encounter within these pages.

1.2 SI units

The need for standardised values and measurements for use in science and engineering was agreed in 1960 at the 11th General Conference on Weights and Measures. Since then, most countries in the world have adopted a universal set of units, known as SI (Systeme Internationale) units. Every value is named and given a symbol, as is every measurement. Greek letters, both upper and lower case, are used as well as letters from the Roman alphabet (the alphabet used for the English language). Table 1.1 gives some examples of SI units related to electrical science and technology.

Table 1.1 Examples of SI units

Value	Symbol	Measurement	Symbol
Current	I	amperes	A
Power	P	watts	W
Resistance	R	ohms	Ω
Energy/ Work	W	joules	J
Length	l	metres	m
Magnetic flux	Φ	webers	Wb

1.3 Multiple and submultiple units

There are many examples in practical electrical engineering where the basic units are large and cumbersome. A prefix is given to these values. The most commonly used are:

1000	kilo	k	10^3
1,000,000	mega	M	10^6
1,000,000,000	giga	G	10^9
1,000,000,000,000	tera	T	10^{12}

Examples of their use:

1000 W	= 1 kW
1,000,000 Ω	= 1 MΩ

Submultiples are numbers less than one. Those most commonly used are:

$\frac{1}{1000}$	0.001	milli	m	10^{-3}
$\frac{1}{1,000,000}$	0.000001	micro	μ	10^{-6}
$\frac{1}{1,000,000,000}$	0.000000001	nano	n	10^{-9}
$\frac{1}{1,000,000,000,000}$	0.000000000001	pico	p	10^{-12}

Examples of their use:

0.001 A	= 1 mA
0.00000 F	= 1 μF

Two words of warning are necessary concerning the application of these prefixes.

1. Note the difference between the symbols M and m (one upper case, the other lower case). Using the wrong value will give a wildly inaccurate answer as the difference between them is 1,000,000,000.
2. Always convert a value into its basic SI unit before using it in an equation.

Example 1.1
Find the current for a 3 kW, 230 V load.

$$I = \frac{P}{V}$$

You must convert 3 kW to its SI unit of watts (W).

$$3 \text{ kW} = 3000 \text{ W, so } I = \frac{3000}{230} = 13.04 \text{ A}$$

Conversion of multiples and submultiples

Two simple rules for converting multiples and submultiples are as follows. When converting large values to small, e.g. kW to W, *multiply* by the multiple.

For example, when converting 3 MΩ to Ω

$$3 \times 1{,}000{,}000 = 3{,}000{,}000 \ \Omega$$

When converting small values to large, e.g. mA to A, *divide* by the multiple. For example, when converting 650 mV to V

$$\frac{650}{1000} = 0.65 \ V$$

Example 1.2
Convert the following.

(a) 4.4 kJ to J
(b) 3.56 MΩ to Ω
(c) 407 W to kW
(d) 1280 μF to F

(a) 4.4 kJ to J:

$$k = kilo : 1000 \quad 4.4 \times 1000 = 4400 \ J$$

(b) 3.56 MΩ to Ω:

$$M = mega : 1{,}000{,}000 \quad 3.56 \times 1{,}000{,}000 = 3{,}560{,}000 \ \Omega$$

(c) 407 W to kW:

$$k = kilo : 1000 \frac{404}{1000} = 0.404 \ kW$$

(d) 1280 μF to F:

$$\mu = micro : \frac{1}{1{,}000{,}000} \quad \frac{1280}{1{,}000{,}000} = 0.00128 \ F$$

Conversion of multiples and submultiples for area and volume

While the principles of conversion between multiples and submultiples remain the same when calculating areas and volumes, the values do not. Be careful, because it is easy to forget and be caught out when carrying out these conversions.

Area

The area of a surface is calculated by multiplying two adjacent sides. For example, a surface of 2 m width and 4 m length will be calculated:

$$4 \times 2 = 8 \text{ m}^2$$

Note that the answer is m^2. This means that when converting m^2 to mm^2, the multiplier will be 1000^2, which is 1,000,000, or 10^6.

So 3 m^2 = 3,000,000 mm^2 or 3×10^6 mm^2

Likewise, when converting mm^2 to m^2, the divider will also be 1,000,000, or 10^6. So, to convert 4 mm^2 to m^2, we divide by the multiplier (10^6):

$$\frac{4}{1,000,000} = 0.000004 \text{ m}^2 \text{ or } 4 \times 10^{-6} \text{ m}^2$$

Volume

Volume is calculated from three measurements, the area (length × width) × height. The answer is in m^3. An object with a base area of 12 m^2 and a height of 23 m will have a volume of

$$12 \times 23 = 276 \text{ m}^3$$

To convert to mm^3 to and from m^3, we must cube the multiplier (1000). So, it becomes $1000 \times 1000 \times 1000 = 1000,000,000$.

$$276 \times 1000 \times 10^3 \text{ or } 276 \times 1,000,000,000$$

$$= 276,000,000,000 \text{ mm}^3 \text{ or } 276 \times 10^9 \text{mm}^3$$

Example 1.3
Solve the following:

(a) Convert 43 m^2 to mm^2.
(b) What is the area of a surface if the length is 5 m and the width is 3200 mm?
(c) Convert 8276 mm^3 to m^3.

(a) $43 \times 1,000,000 = 43,000,000$ m^2 or 43×10^6 m^2

(b) $5 \times \frac{3200}{1000} = 5 \times 3.2 = 16$ m^2

(c) $\frac{8276}{1,000,000,000} = 0.000000008276 \text{ m}^3 \text{ or } 8276 \times 10^{-9} \text{ m}^3$

1.4   Transposing formulae

Most formulae used in science and engineering have a default setting. For example, Ohm's law is usually presented as:

$$R = \frac{V}{I}$$

Resistance, R, equals voltage, V, divided by current, I.

However, the resistance and voltage might already be known, and it is the *current* that needs to be calculated. In this case, the formula must be transposed.

Bishops and draughts

When the formula consists of multiplication and division, a quick way to remember how to transpose is that each component can be moved diagonally across the equals sign from one side to the other, like a bishop in chess, or a draught on a draught board. In this case, we need to make I the subject of the Ohm's Law formula. So, using the bishop in chess analogy:

The Ohm's Law formula $R = \dfrac{V}{I}$

1. I can be moved diagonally across the equals sign and placed next to the R:

   $$RI = V$$

2. R can be moved diagonally down across the equals sign and placed beneath the V:

   $$I = \frac{V}{R}$$

   If we need to find the voltage because resistance and current are known, I is moved diagonally across the equals sign and placed next to the R

   $$RI = V$$

   Although mathematically it makes no difference, it is conventional to place the subject of the formula on the left so once we have transposed the formula we may need to flip it round:

   $$V = IR$$

Reversed function

If the chess analogy is not helpful, then approach the transposition mathematically. Looking at our Ohm's Law formula,

$$R = \frac{V}{I} \text{ can also be expressed as } R = V \div I$$

We wish to make current, I, the new subject of the formula. As it stands, I is a divider; it is divided into voltage. By moving it across the equals sign, it becomes a multiplier:

$$R \times I = V$$

The reverse is true when we want to move the resistance R to leave the current as the subject. R is a multiplier. Moving it across the equals sign will make it a divider:

$$I = V \div R \text{ can also be expressed as } I = \frac{V}{R}$$

Addition and subtraction

Formula that includes addition and subtraction can also be transposed. Any component can be moved across the equals sign; however, if it is a positive (plus) number, it will become negative (minus), and if a negative, it will become a positive. For example, the impedance formula:

$$Z^2 = R^2 + X^2$$

If we wish to make X^2 the subject, we can move the R^2 across the equals sign. It will change from a positive to a negative.

$$Z^2 - R^2 = X^2$$

Again, the convention is for the subject to be on the left.

$$X^2 = Z^2 - R^2$$

Example 1.4
Transpose the following formula to make the bold component into the new subject.

(a) $P = V\mathbf{I}$

(b) $X_L = 2\pi F\mathbf{L}$

(c) $Z^2 = \mathbf{R}^2 + X^2$

(d) $\frac{Vp}{Vs} = \frac{Is}{\mathbf{Ip}}$

(a) $P = V\mathbf{I}$

$$\frac{P}{I} = V; \text{ conventionally } V = \frac{P}{I}$$

(b) $X_L = 2\pi FL$

$$\frac{X_L}{2\pi F} = L; \text{ conventionally } L = \frac{X_L}{2\pi F}$$

(c) $Z^2 = R^2 + X^2$

$$Z^2 - X^2 = R^2; \text{ conventionally } R^2 = Z^2 - X^2$$

(d) $\frac{Vp}{Vs} = \frac{Is}{Ip}$

$$\frac{VpIp}{Vs} = Is \quad VpIp = VsIs \quad Ip = \frac{VsIs}{Vp}$$

1.5 Percentages and efficiency

One percent, 1%, is one-hundredth part of a whole amount.

For example, 1% of $20 = \dfrac{20}{100} = 0.2$

To find a percentage of an amount, the amount is multiplied by the percentage figure divided by 100.

For example find 18% of $35 = \dfrac{18}{100} \times 35$ or $0.18 \times 35 = 6.3$

To convert an amount into a percentage of the whole, divide the amount by the whole and then multiply the answer by 100.

For example, 12 students out of 19 passed an exam.

$$\frac{12}{19} \times 100 = 63.12\%$$

Example 1.5

What is:

(a) 57% of 109
(b) 0.3% of 68
(c) 12.5% of 281

(a) 57% of $109 = \dfrac{57}{100} \times 109$ or $0.57 \times 109 = 62.13$

(b) 0.3% of $68 = \dfrac{0.3}{100} \times 68$ or $0.003 \times 68 = 0.204$

(c) 12.5% of $281 = \dfrac{12.5}{100} \times 281$ or $0.125 \times 281 = 35.13$

Example 1.6
A batch of 750 socket outlets rated at 13 A was issued from the manufacturer, but 420 of them were faulty. What percentage of the socket outlets were faulty?

$$\frac{420}{750} \times 100 = 56\%$$

Efficiency

Efficiency, the percentage difference between input and output, is an important part of electrical science and technology; the efficiency of pumps, transformers and electric motors needs to be calculated as part of electrical design. For example, cable selection depends on the amount of current required by a load. This, in turn, depends on the power dissipated to make the load work.

To calculate efficiency (η) of an electrical load:

$$\eta = \frac{\text{Output power}}{\text{Input power}} \times 100$$

Example 1.7
An electric motor requires 452 W to produce 320 W of power. How efficient is this motor?

$$\eta = \frac{\text{Output}}{\text{Input}} \times 100 \qquad \eta = \frac{320}{452} \times 100 = 70.8\%$$

Example 1.8
A 390 W pump is 83% efficient. How much actual power needs to be dissipated to operate the pump.

$$\eta = \frac{\text{Output}}{\text{Input}} \times 100$$

Transpose to make input power the subject and convert % into a whole number.

$$\frac{83}{100} = 0.83$$

$$\text{Input power} = \frac{\text{Output power}}{\eta} = \frac{390}{0.83} = 469.88 \text{ W}$$

1.6   Trigonometry

Certain calculations for AC supplies and for light levels require a knowledge of basic trigonometry. So, here is a reminder of the main rules for right-angle triangles (Figure 1.1).

Pythagoras' theorem

The longest side of a right-angle triangle is the hypotenuse. It is calculated using Pythagoras' theorem –'The square of the hypotenuse equals the sum of the squares of the other two sides'.

Shown as an equation, and referring to Figure 1.2, Pythagoras' theorem is:

Hypotenuse$^2 = A^2 + B^2$

It can also be presented as:

Hypotenuse $= \sqrt{A^2 + B^2}$

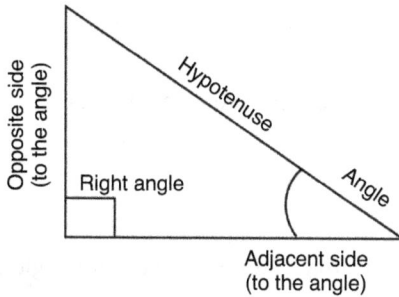

Figure 1.1   The components of a right-angle triangle

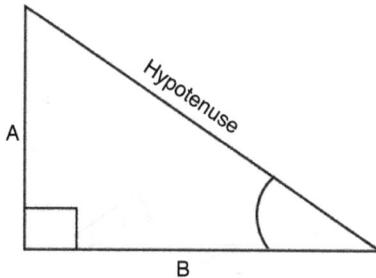

Figure 1.2   Right-angle triangle, stripped down for Pythagoras' theorem

The 3–4–5 triangle

The 3–4–5 triangle is a phenomenon in which a right-angle triangle with an A and B side of 3 and 4 will have a hypotenuse of 5. This also applies to multiples of these numbers, providing the multiples are the same for each side, for example:,

- x 3: 9–12–15
- x 8: 24–32–40

Tangent, sine and cosine

Tangent, sines and cosines can be used to determine the lengths of a triangle's sides as well as its angles.

The abbreviation SOH CAH TOA can be used as a memory aid for the three formulae:

$$\text{Sine} = \frac{\text{Opposite}}{\text{Hypotenuse}} \qquad \text{Cosine} = \frac{\text{Adjacent}}{\text{Hypotenuse}} \qquad \text{Tangent} = \frac{\text{Opposite}}{\text{Adjacent}}$$

To convert an angle into a tangent, sine or cosine on a calculator:

Tan – angle = Tan of the angle
Sin – angle = Sin of the angle
Cos – angle = Cos of the angle

To convert a tangent, sine or cosine into an angle it is:

SHIFT – Tan = angle
SHIFT – Sin = angle
SHIFT – Cos = angle

The same applies for converting sines or cosines into angles.

Example 1.9
Find the cosine, tangent or sine, and the angle ϕ for these triangles (Figure 1.3).

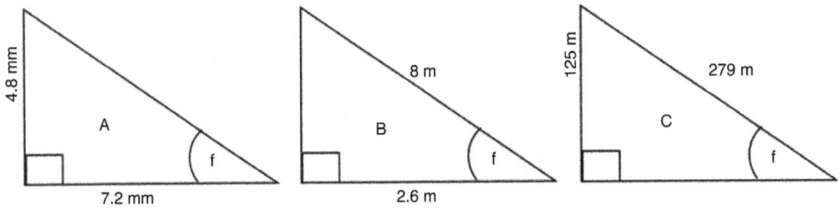

Figure 1.3   Triangles for Example 1.9

TRIANGLE A $\text{Tangent} = \dfrac{\text{Opposite}}{\text{Adjacent}} = \dfrac{4.8}{7.2} = 0.67$ Angle $= 33.83°$

TRIANGLE B $\text{Cosine} = \dfrac{\text{Adjacent}}{\text{Hypotenuse}} = \dfrac{2.6}{8} = 0.33$ Angle $= 71.03°$

TRIANGLE C $\text{Sine} = \dfrac{\text{Opposite}}{\text{Hypotenuse}} = \dfrac{125}{279} = 0.45$ Angle $= 26.62°$

Example 1.10

TRIANGLE A
Find the hypotenuse – we have an angle and the adjacent length so cosine can be used (Figure 1.4):

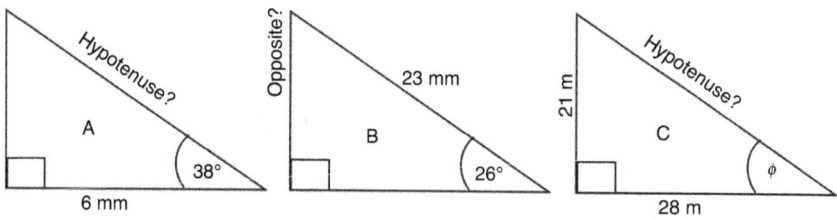

Figure 1.4   Triangles for Example 1.10

$$\text{Cosine} = \frac{\text{Adjacent}}{\text{Hypotenuse}}$$

Transpose to make hypotenuse the subject

$$\text{Hypotenuse} = \frac{\text{Adjacent}}{\text{Cosine}}$$

Cosine of $38° = 0.79$ $\text{Hypotenuse} = \dfrac{6}{0.79} = 7.59$ mm

TRIANGLE B

Find the opposite – we have an angle and the hypotenuse, so sine can be used:

$$\text{Sine} = \frac{\text{Opposite}}{\text{Hypotenuse}}$$

Transpose to make opposite the subject

Opposite = Hypotenuse × Sine

Sine of 26° = 0.44 Opposite = 23 × 0.44 = 10.12 mm

TRIANGLE C

Find the hypotenuse. We have both opposite and adjacent. Both are divisible by 7:

$$28 = \frac{4}{7} \quad 21 = \frac{3}{7},$$

which means that this is a 2–4–5 triangle, so 7 × 5 = 35 m

1.7   Circles

Pi

Pi, symbol π, is a value used to calculate various dimensions of a circle. It is derived from the ratio between circumference and the diameter. This applies to any circle. Pi has infinite decimal places and is usually rounded up to 3.14 or 3.142.

$$\pi = \frac{\text{circumference}}{\text{diameter}} \quad \frac{C}{d}$$

Circumference

The distance around the outside edge of a circle. The formula for circumference is:

$$C = \pi d \quad C = 2\pi r$$

Radius

The radius of a circle is the distance from its centre to the circumference. The formulae for radius are:

$$r = \frac{d}{2} \quad r = \frac{C}{2\pi}$$

$$r = \sqrt{\left(\frac{a}{\pi}\right)}$$

Area

The formulae for the area of a circle is:

$$a = \pi r_2$$

Radians

A radian is defined as the angle subtended at the centre of a circle by a section of the circumference of equal length to its radius (Figure 1.5).

Since the circumference of a circle is $2\pi \times$ radius, there are 2π radians in 360°, so:

$$1 \text{ radian} = \frac{360°}{2\pi} = 57.3° \text{ approximately.}$$

Example 1.11
A circle has a diameter of 296 mm. Find:

(a) circumference
(b) radius
(c) area

(a) $C = \pi d = \pi \times 296 = 930.03$ mm

(b) $r = \frac{d}{2} = \frac{296}{2} = 146$ mm

(c) $a = \pi r^2 = \pi \times 146^2 = 68,822.37$ mm² or 0.069 m²

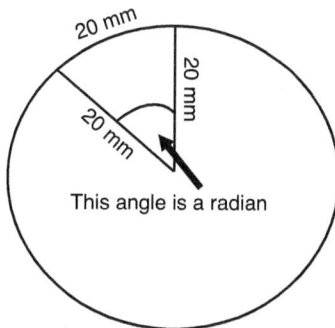

Figure 1.5 The radian of a circle

1.8   Ratios

A ratio is the comparison of one part of a total value to another part. Ratios are used in electrical science and technology to calculate transformer values.

If there are two boxes and box A holds 15 objects, while box B holds 5, then there are three more objects in Box A to every 1 object in Box B. We know this because:

$$\frac{15}{5} = 3$$

The ratio is, therefore, 3:1.

Example 1.12
A qualified electrician fitted 14 lights in an afternoon, while the apprentice fitted 5. What is the ratio of lights fitted between the apprentice and the electrician?

$$\frac{14}{5} = 2.8$$

This means that the electrician fitted 2.8 lights in the time taken for the apprentice to fit one. The ratio is, therefore, 1:2.8.

Example 1.13
A wholesaler sells isolators manufactured both by Company A and Company B. The Company A isolators outsell Company B's by a ratio of 24:8. If 174 isolators are sold in a week, how many of those are Company A types and how many Company B types?

Simplify ratio $\frac{24:8}{4} = 6:2$, which can be further simplified to 3:1. This means that 3 Company A isolators are sold for every single Company B type. In other words, Company B sells $\frac{1}{3}$ of Company A's total.

Company B sells $\frac{174}{3} = 58$ isolators

Company A sells $174 - 58 = 116$ isolators

1.9   Reciprocation

To reciprocate a number means to invert it. This is used when calculating the equivalent value of resistances connected in parallel, and capacitors connected in series.

Every whole number can be expressed as:

$\frac{\text{Number}}{1}$ Reciprocated, this becomes $\frac{1}{\text{Number}}$. This is also expressed as Number^{-1}.

In essence, all numbers are fractions and, when reciprocated, can be calculated as such by finding a common denominator. However, modern scientific calculators include a reciprocation button, labelled X^{-1}.

So, to reciprocate a number using the calculator:

$$\text{Number} - \text{then} - X^{-1} =$$

This means that you can quickly add a string of reciprocated numbers, such as those below:

$$6^{-1} + 4^{-1} + 18^{-1} = 0.47$$

The answer is also a reciprocated number. To correct this, press X^{-1} once more:

$$0.47^{-1} = 2.12$$

Example 1.14
Add the following values, using the reciprocation function on your calculator:

$$23.8 + 52.4 + 13 = 23.8^{-1} + 52.4^{-1} + 13^{-1} = 0.14^{-1} = 7.25$$

Note that the answer to a reciprocated string of numbers will always be a lower value than the lowest number in the string.

1.10   Exercises

1. What is an SI unit?
2. Convert the following:
   (a) 3267 Ω to MΩ
   (b) 34 μF to F
   (c) 0.56 kV to V
   (d) 0.06 m^2 to mm^2

3. Transpose the following formula to make the bold component into the subject:
   (a) $P = V\boldsymbol{I}$

   (b) $XL = 2\pi F\boldsymbol{L}$

   (c) $Xc = \frac{1}{2\pi F\boldsymbol{C}}$

   (d) % slip $= \frac{ns - \boldsymbol{nr}}{ns}$

4. A total of 28 switches out of 144 were found to be faulty. What percentage is this?

5. In a class, 68% of students out of 22 failed to understand how to calculate percentages, how many students was this?

6. In a factory, 14% of 150 m drum of armoured cable was left over after being used as a sub-main. What was the length of the cable run?

7. A 509 W electric pump required 798 W to operate. How efficient was the pump?

8. What is the supply power dissipated by a 12.7 W electric motor if it is 70% efficient?

9. Supply the formulae for each of these:
    (a) Cosine =
    (b) Sine =
    (c) Tangent =

10. What is the angle of a right-angled triangle if its adjacent side is 5.3 m and its opposite is 1.6 m?

11. What is the cosine of a triangle which has an opposite length of 26 mm and an adjacent side of 33 mm?

12. What is the hypotenuse of a triangle if its other two sides are 7 mm and 2 mm, respectively?

13. What is the length of the opposite side if a right-angled triangle has an angle of 66° and an adjacent side of 33 mm?

14. A circle has a circumference of 7.2 m. Find:
    (a) Diameter
    (b) Radius
    (c) Area

15. Describe the radian of a circle.

16. Wholesaler Green's sold 558 items of stock in a day, Wholesaler Yellow, sold 1008 items. By what ratio did Wholesaler Yellow outsell Wholesaler Green?

17. If an electrician installed her share of 81 socket outlets, at a ratio of 9:1, how many sockets did her apprentice install?

18. What does reciprocation mean in mathematics?

19. The following resistors were connected in parallel. What is their equivalent resistance?

    46 Ω, 69 Ω, 21 Ω and 91 Ω

20. The following capacitors were connected in series. What is their equivalent capacitance?

    883 μF, 698 μF and 229 μF.

1.11   Multiple choice questions

1. What are the SI unit symbols for current?
   (a) V and W
   (b) W and I
   (c) I and A
   (d) R and m

2. What are the SI unit symbols for resistance?
   (a) I and A
   (b) R and X
   (c) V and W
   (d) R and Ω

3. 47 mA is
   (a) 0.047 A
   (b) 47 A
   (c) 4.7 A
   (d) 0.0047 A

4. Mega is:
   (a) 10^6
   (b) 10^9
   (c) 10^3
   (d) 10^4

5. Micro is
   (a) 10^{-3}
   (b) 10^{-4}
   (c) 10^{-6}
   (d) 10^{-9}

6. If a number is a multiplier on one side of the equals sign, what will it be on the other side?
   (a) Minus
   (b) Divider
   (c) Percentage
   (d) Multiplier

7. 83% of 0.6 =
   (a) 0.88
   (b) 0.24
   (c) 0.64
   (d) 0.48

8. 96 out of 113 =
   (a) 77.2%
   (b) 84.96%
   (c) 59.73%
   (d) 94.3%

9. Tangent =
   (a) Opposite over adjacent
   (b) Hypotenuse over opposite
   (c) Adjacent over opposite
   (d) Opposite over hypotenuse

10. Adjacent over hypotenuse =
    (a) Tangent
    (b) Cogent
    (c) Cosine
    (d) Sine

11. What is the ratio if I have £150 and you have £30?
    (a) 15:1
    (b) 20:60
    (c) 9:3
    (d) 5:1

12. Changing 3 to $\frac{1}{3}$ is called:
    (a) Simplification
    (b) Applying inverse square law
    (c) Reciprocation
    (d) Inversely proportionate

Chapter 2

Regulations and safety

2.1   Introduction

Although electricity is our friend and drives much of the modern world, it is also dangerous. It is a form of energy that can both kill, if it passes through the human body, and cause fires through overheating.

The normal user of electricity, whether at home, or at work, must be protected from harm. This is down to the designer, the electrician and electrical engineer. There must be no possibility that anyone can touch a live part, and when a fault occurs, the supply must be shut down before the user is injured. This is down to correct protection being included in the design of an installation, a subject explored in detail in Chapter 14 of this book.

The electrician is also at risk when working on electrical systems, particularly those already in place and that have the potential to become live if sufficient precautions are not carried out. There are a number of procedures, regulations and guidance notes intended to protect the electrician and others working in their vicinity, such as on a construction site.

2.2   Health and safety at work

In the United Kingdom, health and safety in the workplace is overseen by the Health and Safety Executive (HSE), who have power to enforce regulations and bring prosecutions in necessary. They are also an advisory organisation who can give assistance and advice.

The overarching set of **statutory regulations** in the United Kingdom is the *Health and Safety at Work Act*. While not specific about exactly how health and safety measures are carried out, this document states the need for:

- Health and safety to be an integral part of any workplace
- Specific health and safety roles to be assigned to personnel within an organisation
- Awareness that responsibility for health and safety begins at the top of an organisation, but is the responsibility of all employees and also of visitors to a workplace
- Knowledge of health and safety by all members of an organisation

More detailed and specific requirements for working safely and healthily are found in the subject-related statutory regulations, such as the Working at Height, and the Personal Protective Equipment Regulations. These are underpinned by **non-statutory** codes of practice and guidance notes. These are practical documents that explain how health and safety regulations can be applied in the workplace.

KEY TERM: Statutory Regulations – These are regulations that are law and to which you must conform. Any breach of statutory regulations could end up as a court case, and punished by fines and possible imprisonment.

KEY TERM: Non-Statutory – These are not backed directly by law but are practical and detailed guides on how to apply statutory regulations. This does not mean that following their requirements is optional. Failure to adhere to the requirements of a non-statutory code of practice or guidance note is, by default, a failure to adhere to the requirements of the statutory document that backs it up.

Risk assessment

The second layer of statutory requirements are the Management of Health and Safety at Work Regulations. These focus particularly on risk assessment.

Risk assessment is something we all carry out in our daily lives, for example, before crossing a road. It is a natural act intended to keep us safe when approaching a hazardous action. In industry and within an organisation, there is a requirement to write down formal risk assessments to warn of hazards and risk and instruct people on how to avoid harm. While risk assessments can be tailored to suit an individual situation or procedure, the main headings will be the same.

Hazard

Something that has the potential to cause harm. Working at height for example, or in a confined space.

Risk

The harm that could come from that hazard. Take working at height, the risks are:

- Falls
- Dropping objects onto people below

Protective actions

How the risk is reduced. For working at height this could be:

- Work from a mobile tower and not a ladder
- Guard rails on the tower
- Kick boards to prevent objects falling
- Isolate the area around the base of the tower to keep people away

Review date

Risk assessments may apply to a specific stage in a project and once that work has been completed, the risk assessment is no longer valid, or the organisation may change in some way. This means that the risk assessment must be revised to make it relevant to the new circumstances.

Other fields may be added in, according to the purpose of the risk assessment. For example:

- Who will be harmed
- Level of harm

Method statements

Other documents and procedures are also employed to keep the workplace as safe as possible, include the method statement which is, essentially, a step-by-step guide to how a project is to be carried out safely and efficiently.

Permit to work

To be completed and applied when carrying out a particularly hazardous task, e.g.

- Working in a confined space
- Working near active machinery
- Working on live electrical systems

A permit to work should state the start and finish times for the task and be signed off by a senior, experienced person who understands the work and its risks. It should include a method statement and risk assessment. The finish time will, of course, be an estimate, but it is a guide as to when the person doing the work should return from the job. If the finish time is exceeded, it acts a prompt for the supervisor to check up and make sure no accident has occurred.

2.3   Electrical regulations

Statutory electrical regulations

Electricity at Work Regulations

The first level document that applies to electrical safety are the *Electricity at Work Regulations*. These apply to every workplace, office, shop, factory or farm, not just a site where electrical work is being carried out. In principle, they also apply to the home. The *Electricity at Work Regulations* state that:

- No one at work should be at risk of harm from electrical systems, or work taking place on those systems.
- Electrical systems should be adequate and correctly installed.
- Electrical systems should be regularly inspected and tested.
- There should be protection in place, against shock and fire.

- No one should work on live systems unless it is absolutely necessary, an example of this being testing and fault diagnosis.

Electricity safety, quality and continuity regulations

These regulations are concerned with the supply of electricity to an installation. They are aimed at **Duty Holders** for the design, installation and maintenance of electricity supplies. This includes the system by which electricity is distributed by the distribution network operator (DNO) and where it connects to a premises. The regulations include the need to provide:

- A continuous and consistent supply
- Protection from overcurrent, and voltage disturbances
- Earthing and connections for earthing
- Substation safety

Safe installation and placement of overhead and underground cables are also included in these regulations as well as safe and correct generator installation and connection.

KEY TERM: Duty Holder – The person responsible for a building, service or process.

The Building Regulations

The building regulations apply to the construction of residential premises, such as houses and flats. They are divided into a number of parts, each focussed on a specific area of construction. Although part P deals with electrical installation in residential properties, many of the other parts are relevant.

For instance, take parts A and B, which require any penetration of a wall, by a cable, containment, duct or pipe, to be made good around the penetration to the same standard as the rest of the wall, ceiling or floor. It is also important that damp, sound (part E) or fire cannot spread via that penetration.

The building regulations, particularly part P, do not overrule the various sets of electrical regulations, rather they concentrate on the points that apply specifically to residential dwellings.

Parts of the building regulations

Part A: Structure
Part B: Fire safety
Part C: Site preparation and resistance to contaminates and moisture
Part D: Toxic substances
Part E: Resistance to sound
Part F: Ventilation
Part G: Sanitation, hot water safety and water efficiency
Part H: Drainage and waste disposal
Part J: Combustion appliances and fuel storage systems
Part K: Protection from falling, collision and impact
Part L: Conservation of fuel and power

Part M: Access to and use of buildings
Part O: Overheating
Part P: Electrical safety
Part Q: Security in dwellings
Part R: Infrastructure for electronic communications
Part S: Infrastructure for charging electric vehicles
Part T: Toilet accommodation
Part 7: Material and workmanship

Non-statutory electrical requirements

BS7671 IET *Wiring Regulations*

Although they are called *Wiring Regulations, BS7671* is a non-statutory document that gives detailed requirements for safe and adequate electrical installations. It is divided into parts which deal with subjects including installation or cables and equipment, protection from shock and overcurrent, and inspection and testing. The second section of the book consists of appendices, which provide supporting information for the regulations which aids compliance by the electrician and designer.

IET *On-Site Guide*

Aimed mostly at residential installations, but relevant to most others, the *On-Site Guide* both supplements *BS7671*, by providing sets of tables not available in *BS7671*, and offers practical guidance on how to apply the regulations.

IET guidance notes

A set of practical manuals which give detailed guidance on such subjects as protection from shock, earthing, and inspection and testing. The guidance notes tie-in closely with *BS7671* and the *On-Site Guide*.

GS.38 electrical test equipment for use by electricians

An HSE practical guide to safe use of electrical test instruments. Among its requirements are:

- Finger shields for test probes
- Maximum 4 mm of metal tip exposed at the end of a probe
- Leads to be flexible and colour-coded
- All test instruments to be calibrated annually

2.4   Safe working practices

Electric shock

An electric shock is the passage of an electric current through the body. The amount of current which is lethal varies from person to person, and also depends on the parts of the body in which it flows. To understand why we are 'shocked', we must realise that every movement we make, conscious or unconscious, is produced by muscles reacting to minute electric currents generated in the brain. These currents are distributed to the correct muscles by the 'conductors' of the nervous system.

If a current much larger than the one usually carried is forced through the nervous system, the muscles react much more violently than normal, and hence we experience the 'kick' associated with an electric shock. If, in addition, the nerves carry the excess current to the brain, it may destroy or cause temporary paralysis of the cells that generate the normal currents. Destruction of these cells means almost instant death as the heart muscles cease to operate and no blood is circulated. Paralysis results in unconsciousness but if the lung muscles are not operating, death from suffocation will follow in a few minutes.

Severity of shock

The severity of shock depends on the amount of current flowing in the body, the path it takes and the time for which it flows. As seen in Chapter 3, the amount of current flowing increases as the applied voltage increases, and decreases as circuit resistance increases. This means that in identical circumstances, a worse shock will be received from a high voltage than from a low voltage.

For a given voltage, the severity of the shock received will depend on circuit resistance, which is made up of the following parts.

- Resistance of the installation conductors. These form such a small proportion of the total resistance that usually they may be ignored.
- Resistance of the body. This varies considerably from person to person and, for a particular person, with time. As the body itself is made mainly of water, its resistance is quite low, but it is covered by layers of skin which have high resistance.

The main resistance variations occur in the skin. Some people have a naturally hard skin which has high resistance, and others have soft, moist skin of low resistance. If the skin is wet, the moisture penetrates the pores, giving paths of low resistance. A natural result of electric shock is for the victim to perspire profusely, which reduces body resistance and increases the severity of the shock.

Contact with the general mass of earth

The body is normally separated from the conducting mass of earth by one or more layers of insulating material; that is, shoes, floor coverings, floors and so on. It is the resistance of these insulators which normally prevents a shock from being serious. For instance, a person wearing rubber-soled shoes standing on a thick carpet over a dry wood floor can touch a live conductor and feel nothing more than a slight tingle. The same person standing on a wet concrete floor in his bare feet would probably not live to describe his sensations on touching it!

This account is oversimplified, but gives an indication of what occurs. Although data on the effects of shock are available (BS PD 6519), it would be unwise to assume that a given shock current flowing for a given time would always have certain effects or would be safe, because shock effects appear to differ from person to person and are affected by factors such as ambient temperature, health and so on. It is always best to ensure as far as possible that there will be no shock by scrupulously following *BS 7671*. An effective backup against shock to earth is to fit a residual current device with a rating not exceeding 30 mA (see Chapter 14).

In the event of shock

As explained in the previous section, severe electric shock is often accompanied by a form of paralysis of the nervous system. The heart continues to beat, and the victim is not yet dead, but as breathing has stopped, the victim will soon die if no action is taken.

Artificial respiration keeps the patient's lungs working until the body's own system can recover and take over. It is very simple to apply and is basically the same as that used in drowning cases. Every person should be proficient in its use.

Precautions

If you witness an accident caused by electric shock resulting in unconsciousness or find an unconscious colleague, remember the following points.

- Ensure that the patient is not still in contact with the electrical system. If they are, and you touch the body, you may receive a shock too. Switch off the supply, or if this is not possible, drag the patient clear with dry clothing or some other insulator.
- Do not waste time trying to find out if the patient is still alive. It is vital to start respiration at once.
- Summon assistance, namely doctor, ambulance and so on, at once, but do not delay artificial respiration to do so. Most people own a mobile device so an emergency call can made easily. If you are on your own, shout for help periodically.

Mouth-to-mouth artificial respiration

Sometimes called the kiss of life, appears to be accepted generally as the quickest and most effective system.

1. The patient should be quickly laid on his or her back, with the head tilted as far backwards as possible. This will open and straighten the air passages.
2. Clear the mouth of foreign objects such as false teeth, vomit, and so on, and make sure that the tongue is not blocking the airway.
3. Close the patient's nose by pinching the nostrils, place your mouth firmly over the patient's and blow.
4. When the chest is inflated, remove your mouth, drawing in breath as you do so. The patient's chest will deflate, after which the cycle should be repeated. The process should take place 10–12 times each minute; that is, once every five or six seconds. Counting the seconds to yourself may help you to prevent slowing down, which can prove fatal.

An alternative method is mouth-to-nose artificial respiration, which is easier in some cases.

1. The victim's mouth should be closed with the thumb, and your mouth placed firmly over the patient's nose for the blowing operation.
2. If the chest is not inflated, check the head position and try again.
3. Failure on the second attempt means that you must change to the mouth-to-mouth system at once.

Please note that professional opinion on best practice in first aid does change periodically and an appropriate first aid course is the best way to receive the most up-to-date advice.

Safe isolation

The Electricity at Work Regulations states that we should never work on a live circuit unless it is unavoidable. In such cases, work must be supervised by a fully qualified person who puts the permit to work system into effect.

So, most electrical work should be carried out with the circuit or installation de-energised. Even then, there is a risk that someone might misunderstand why the power is off and attempt to switch it back on. We also need to be certain that the circuit really is dead, before starting work.

The safe isolation procedure has been developed to minimise the risk of shock when working on a circuit believed to be dead. Figure 2.1 shows an example of a full safe isolation kit.

The safe isolation procedure

1. Identify source of the circuit and isolate. Make safe by locking circuit breakers in the off position or taking fuses with you. Put up warning notices and erect a barrier around the consumer unit or distribution board.
2. Select an approved voltage indicator. This cannot be a cheap testing pen or neon screwdriver. This must be an instrument with two probes and that conforms with BS EN 61243-3.
3. Prove the indicator is working correctly on either a known live supply or using a proving unit.
4. Test the circuit.
   For single phase, test between
   - (i) line and neutral
   - (ii) line and earth
   - (iii) neutral and earth

5. For three phase
   - (i) L1 and L2
   - (ii) L1 and L3
   - (iii) L2 and L3
   - (iv) L1 and neutral

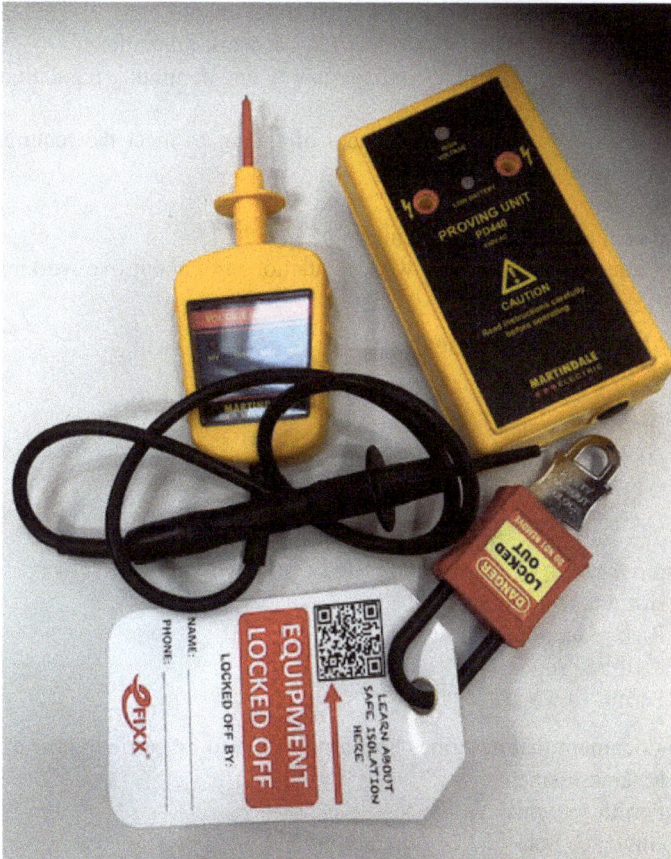

Figure 2.1 Safe isolation kit

(v)   L2 and neutral
(vi)   L3 and neutral
(vii) L1 and earth
(viii) L2 and earth
(ix)   L3 and earth
(x)   Neutral and earth

6.   Re-prove the voltage indicator either on a known live supply or a proving unit.
7.   Begin work.

2.5   Exercises

1.   Which Act provides the overarching set of health and safety regulations?
2.   What is the difference between a statutory and non-statutory document?
3.   What organisation is responsible for the implementation and monitoring of health and safety in the United Kingdom?

4. What are the three main parts of a risk assessment?
5. Give two requirements of the electricity at work regulations
6. To whom are the electricity, safety, quality and continuity regulations mainly intended for?
7. What are the possible consequences of failing to meet the requirements of statutory safety regulations?
8. What is *BS7671*?
9. What code of practice supplements *BS7671*?
10. What does the electricity at work regulations say about live working?

2.6   Multiple-choice questions

1. What is a DNO?
   (a) Distribution neutral overcurrent
   (b) Distribution network operator
   (c) Dielectric network optimisation
   (d) Dielectric neutral overcurrent

2. What document should be completed if live working is necessary?
   (a) Hot works certificate
   (b) Risk reduction request
   (c) Method proposal
   (d) Permit to work

3. The document that describes how a project is to be carried out is a:
   (a) Risk assessment
   (b) Permit to work
   (c) Guidance note
   (d) Method statement

4. Who is the person responsible for a building area or process?
   (a) Duty holder
   (b) Health and safety manager
   (c) Skilled person
   (d) Key holder

5. Which part of the building regulations applies to the electrical installation in a residential property?
   (a) A
   (b) E
   (c) S
   (d) P

6. What are the two main types of artificial respiration?
   (a) Holger Neilson and mouth-to-mouth
   (b) Cardiac massage and mouth-to-mouth

(c) Heimlich manoeuvre

(d) Cardiac massage and shock treatment

7. When should a voltage indicator be proved to be working correctly during safe isolation?

(a) Never, it should be calibrated

(b) Before and after use

(c) Before use only

(d) Every three months

8. Which conductors should be tested during the safe isolation procedure on a single-phase circuit?

(a) L–N, N–E, L–L

(b) L–N, L–E, N–E

(c) L–N, N–L, N–E

(d) L–E, N–E, E–L

9. Which code of practice applies to test instruments?

(a) GS.38

(b) Guidance Note 1

(c) IET *On-Site Guide*

(d) Building regulations part E

10. What should be carried out on test instruments once-a-year?

(a) Moving coil deflection check

(b) Calibration

(c) Visual inspection

(d) Registration

Chapter 3

Basic electrical units and circuits

3.1   Simple electron theory

Electricity in the form of lightning must have been apparent to humankind from their earliest cave-dwelling days. It has taken centuries for us to understand the nature and usefulness of electricity: a startling pyrotechnic novelty in the hands of showmen, the subject of serious investigation and discoveries, and, finally, at the end of the nineteenth century, successfully harnessed to provide light, heat and mechanical power. Electricity is now a universal source of energy that powers the modern world in ways those early dabblers and scientists could never have imagined. With this, increased use has come a greater awareness of its nature. Present-day understanding is based on the theory of atomic structure, although our knowledge is still far from complete.

The atom

All matter is composed of atoms, which often arrange themselves into groups called molecules. An atom is so small that our minds are unable to appreciate what vast numbers make up even a very small piece of material. Eight million atoms, placed end-to-end, would stretch for about 1 mm.

The atom itself is not solid but composed of even smaller particles, separated from each other by space. These are:

- Nucleus – the centre of each atom, made up of protons and neutrons.
- Neutrons – particles that have no net electrical charge.
- Protons – said to have a positive charge.
- Electrons – which complete the atom and are in a constant state of motion, circling the nucleus in the same way a satellite circles the Earth. Each electron has a negative charge.

Atoms of different materials can be identified by the numbers of their electrons. In the complete state, each atom has equal numbers of protons and electrons, so that positive and negative charges cancel out to leave the atom electrically neutral.

The atoms in solids and liquids are much closer together than those in gases, and in solids they are held in a definite pattern for a given material. Where are more than two electrons in an atom, their paths of motion are in **shells** and orbit in all directions around the nucleus.

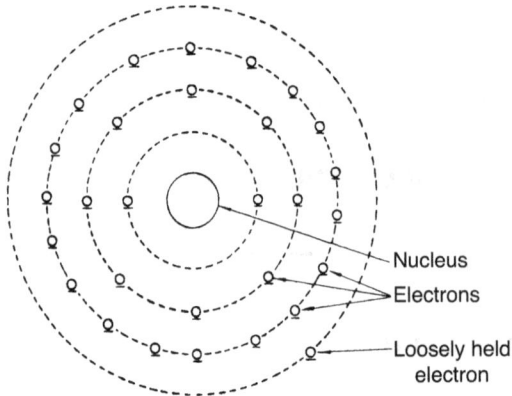

Figure 3.1   Simplified representation of a copper atom

An atom is three-dimensional, and not two-dimensional as indicated in the simplified form of Figure 3.1. This shows a simple representation of a copper atom, which has 29 electrons and 29 protons. Their paths are, in fact, **elliptical** rather than circular.

An electron in the outer shell is held in position weakly and often breaks free, moving at random among the other copper atoms. An atom that has lost an electron in this way is left with an overall positive charge, since it now has one more positive proton than negative electrons. Such an incomplete atom is called a positive ion.

KEY TERM: Shell – A collection of electrons orbiting the nucleus is a way that represents a shell of particles.

KEY TERM: Elliptical – An orbit that is not circular but elongated into an oval shape.

The movement of free electrons in a conductor depends on the laws of electric charge. These are:

- Like charges repel
- Unlike charges attract.

Electric current (symbol, I; measurement, ampere A (more generally, this is shortened to amp))

Figure 3.2(a) represents a block of **conducting material** that contains free electrons moving at random among positive ions. If a battery were connected across the block as shown in Figure 3.2(b):

- Negatively charged, free electrons close to the positive plate will be attracted to it, since unlike charges attract.
- Free electrons near the negative plate will be repelled from it, and a steady drift of electrons will take place through the material from the negative battery terminal to the positive battery terminal.

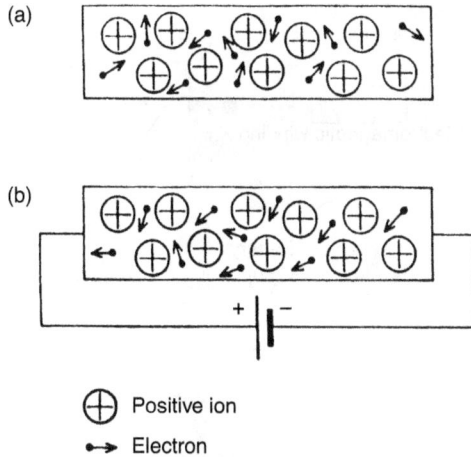

Figure 3.2 Movement of electrons: (a) random movement of electrons in conductor and (b) drift of electrons towards positive plate when battery is connected

For each electron entering the positive terminal, one will be ejected from the negative terminal so that the number of electrons in the material remains constant. Atoms that have become positive ions are unable to move in a solid, which is why they do not drift to the negative terminal.

KEY TERM: Conducting material – A material that allows electricity to flow through it easily.

The rate of movement of electrons through a solid is very low, but, since free electrons throughout the material start to drift immediately when the battery is connected, there is little delay in the *effects* of this movement.

The drift of electrons is called electric current; so, to some extent, we have been able to answer the question 'What is electricity?' However, we have no clear understanding of the nature of an electron, so our knowledge is far from complete. We do know that they:

- Are an elementary particle – not made up from other particles
- Have little mass compared to protons and neutrons
- Have the characteristics of both a particle and a wave

In a battery, electrons are ejected from the negative plate and into the conductor, along which they drift to the battery's positive plate. Once there, they pass through the battery and are ejected once more into the conductor to repeat the process. Thus the electrons circulate but must have a continuous conducting path, or closed circuit. If the circuit is broken, the drift of electrons will cease immediately.

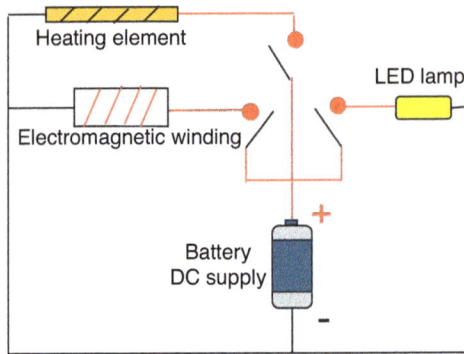

*Figure 3.3   Circuit to illustrate heating, magnetic and chemical effects of electric
current*

Current direction

We have only recently come to understand that an electric current consists of a drift of
electrons. Long before this theory was put forward, current was thought of as an 'electric
fluid' which flowed into conductors from the positive plate of a battery to the negative.

This direction of current, called conventional current direction, was thought to
be correct for many years and was the basis for many rules. We now know that this
assumed current direction is incorrect and that current in a solid actually consists of
an electron drift in the opposite direction. Despite this, we continue, by convention,
to indicate current direction external to the source, as being from the positive to the
negative terminal. In most of the applications we shall consider, the actual direction
of the current does not affect the performance of equipment; because of this, we
shall continue to use conventional current direction.

3.2   Electrical charge and unit of current

If we wish to measure a length, we do so with a ruler or tape measure, which is marked
off in specific units. Since an electric current is invisible, we measure it using test
instruments. The operation of many of these instruments depends on the **magnetic
field** set up by a current in a conductor. These instruments are described in Section 3.9.

> **KEY TERM: Magnetic Field – The area around a magnet in which
> magnetic flux is present. It is within this field that magnetic effects are
> observed. A magnetic field is also present in the material of the mag-
> net itself.**

Charge (symbol for charge, Q; measurement, coulombs C)

Electric charge (symbol Q) can also be thought of as a quantity of electricity. It may
seem that the electron itself could be used, but it is far too small for practical
purposes. The unit for charge is the coulomb (symbol C). One coulomb is much

larger than the charge carried by an individual electron. In fact, it equals the combined charge of six million, million, million (6×10^{18}) electrons.

A body with:

- a surplus of electrons is said to be negatively charged,
- a shortage of electrons is positively charged.

Both these amounts of charge could be measured in coulombs, named after Charles Coulomb (1736–1806), a French physicist.

If the drift of electrons in a conductor takes place at a rate of one coulomb per second, the resulting current is one amp. In other words, a current of one amp indicates that charge is being transferred along the conductor at the rate of one coulomb per second; hence:

$$Q = It$$

where Q = charge transferred in coulombs; I = current in amperes; and t = time during which the current flows in seconds.

Calculation for 1 A current flow:

$$Q = It \text{ transpose formula } I = \frac{Q}{t}$$

If 1 C flows for 1 second, then $I = \frac{1}{1} = 1$ A

Example 3.1
If a total charge of 500 C is to be transferred in 20 s, what is the current flow?

$$Q = It \text{ transpose formula } I = \frac{Q}{t} \quad I = \frac{500}{20} = 25 \text{ A}$$

Example 3.2
A current of 12.5 A passed for 120 s. What quantity of electricity is transferred?

$$Q = It = 12.5 \times 120 = 1500 \text{ C}$$

Example 3.3
A current of 0.15 A must transfer a charge of 450 C. For how long must the current pass?

$$Q = It \text{ transpose formula } t = \frac{Q}{I} \quad t = \frac{450}{0.15} = 3000 \text{ s or 50 min}$$

3.3 Effects of electric current

Electrons are far too small to be seen even with the best microscopes available, and the detection of current would be impossible if it did not produce effects that are more easily detected. There are many such effects, but the four most important are (Figure 3.3):

- heat
- chemical
- magnetic
- electronic – semiconductor effects (for the movement of electrons in a semi-conductor, see Chapter 18)

Heat is generated when current flows in a wire. The amount of heat produced in this way depends on a number of factors, which will be considered later, but in essence, the temperature can be controlled by:

- variation of current
- conductor material
- conductor length and **cross-sectional area**.

KEY TERM: Cross-sectional area – The area of the face of a conductor. Cables are sized according to the cross-sectional area of their conductors, e.g. 1.5 mm^2, 2.5 mm^2, etc.

In this way, the conductor can be made red or white hot, as with an electric fire element or filament lamp. It can also be made to carry current and remain relatively cool, as with an electric cable.

When current passes through chemical solutions, it can cause basic changes to take place in them. Examples of this are a battery and electroplating. Some of these chemical effects will be further considered in Chapter 17.

A current flowing in a coil gives rise to a magnetic field, and this principle is the basis of many electrical devices such as the motor, transformer or relay. The magnetic effect is the subject of Chapters 8 and 9.

Figure 3.3 shows a circuit in which the same current passes in turn through:

- an electromagnet (magnetic effect)
- a lead–acid cell (chemical effect)
- a light-emitting diode (LED) lamp (**electroluminescence**)

The heating and magnetic effects will be apparent owing to the heating of the heating element and the attraction of the iron armature. The chemical effect in the battery is demonstrated if the changeover switch is operated. Table 3.1 lists some of the common devices relying on these three electrical effects.

KEY TERM: Electroluminescence – Light that results from the movement of electrons in a semiconductor (see Chapter 18).

Table 3.1   Some devices relying on effects of electrical current

Magnetic effect	Heating effect	Chemical effect
Relay	Filament lamp	Cells and batteries
Induction furnace	Electric heater	Electroplating
Contactor	Electric cooker	Fuel cell
Motor	Electric kettle	
Transformer	Electric iron	
Circuit breaker	Circuit breaker	
Data recorder	Welder	
Ammeter	Furnace	
Generator	Fuse	
Voltmeter		
Induction hob		

3.4   Electric conductors and insulators

Electric conductors

We have already seen that an electric current is the drift of free electrons in a solid. It follows that for a material to be capable of carrying current, its electrons will be loosely held by their atoms. These become detached at normal temperatures or can be detached by the application of an electric charge. These materials are called conductors. A list of conductors, with remarks on their properties and uses, is given in Table 3.2.

Silver is the best electric conductor, but its high cost and poor physical properties prevent its use in cables.

Copper has the next best conducting properties. Its malleability (the ease with which it can be beaten into shape) and its ductility (the ease with which it can be drawn into strands) make it the natural choice as a conductor for cables. Many heavy supply cables, and almost all wiring cables, have copper conductors.

Aluminium is a poorer conductor than copper, but it is lighter and cheaper. Since aluminium is not as flexible as copper, it cannot be drawn thin enough to be formed into fine wires. It also poses connection problems owing to rapid surface corrosion. Aluminium is often used for large **high-voltage** cables because of its lower cost and relative lightness.

KEY TERM: High Voltage

o **High Voltage (HV) – voltages of 1000 V and above**
o **Low Voltage (LV) – voltages between 50 V and 1000 V**
o **Extra low voltage (ELV) – voltages less than 50 V**

Electric insulators

Materials composed of atoms that have tightly bound electrons have no free electrons available to form an electric current, so none can flow. Such materials are

Table 3.2   Electric conductors

Material	Properties	Application
Aluminium	Low cost and weight	HV Power cables
Brass	Easily machined; resists corrosion	Terminals, plug pins
Carbon	Hard; low friction with metals	Machine brushes
Gold	Expensive; does not corrode	Plating on contacts
Iron and steel	Common metal	Conduits, trunking, fuseboard cases, etc. (protective
Lead*	Does not corrode; bends easily	Power cable sheaths (protective conductor)*
Mercury	Liquid at normal temperatures; vaporises and is toxic	Special contacts, discharge lamps
Nickel	Hard; resists corrosion	Heating elements (with chromium)
Silver	Expensive; the best conductor	Fine instrument wires, plating on contacts
Sodium	Vaporises readily	Discharge lamps
Tin*	Resists attack by sulphur	Coating on copper cables
Tungsten	Easily drawn into fine wires	Lamp filaments
Niobium phosphide	Can be produced at extremely small sizes. Outer surface more conductive than centre but an excellent conductor	Fine wiring for electronic circuits
ACCM	A carbon and glass fibre core surrounded by aluminium strands	Lightweight but extremely conductive, used for high-voltage transmission

*These materials are becoming rare as conductors.

called electric insulators. There are very many types of insulating material, but a few of those in common use in the electric industry are listed in Table 3.3.

Table 3.3 is far from complete. For example, many new types of plastic have been developed as cable insulation, each having special properties. For example:

- Polychloroprene (PCP) has particularly good weather-resisting properties.
- Chlorosulphonated polyethylene (CSP) has increased resistance to physical damage.

No material is a perfect insulator, and all will pass a small 'leakage current'. This leakage is usually so small compared with the operating currents of the equipment that it may be ignored in most cases.

Application of conductors and insulators

An electric cable is a very good example of the application of conductors and insulators. Figure 3.4 shows a typical twin-and-earth house-wiring cable with a protective conductor. The current-carrying conductors are insulated with PVC. An outer sheath, also of PVC, keeps the conductors together and protects them from

Table 3.3   Electric insulators

Materials	Properties	Application
Rubber	Flexible; life affected by high temperatures. Will perish eventually.	Flexible cable insulation
Polyvinyl chloride (PVC)	Versatile, hard wearing and flexible	Cable insulation (all types and sizes), also conduit and fittings
Cross-linked polyethylene (XLPE)	Emits little smoke or fumes when burning	Cable insulation (medium and large sizes) where low smoke emission cables are required e.g. public buildings, also FP cable sheaths
Impregnated paper, varnished cambric	Rather stiff, but unaffected by moderate temperatures, hygroscopic	Transformer winding insulation
Magnesium oxide (mineral insulation)	Powder; requires containing sheath; not affected by very high temperatures; very hygroscopic	Mineral-insulated cables
Mica	Not affected by high Temperatures and unaffected by electrical discharge	Electronic components, generators and HV switches
Asbestos*, glass fibre	Reasonably flexible; not affected by high temperatures	Cable insulation in cookers, fires, etc.
Glass**	Rigid; easily cleaned	Overhead-line insulators
Porcelain	Hard and brittle; easily cleaned	High-voltage insulators and in some components
Rigid plastics	Not as expensive as porcelain and less brittle	Fuse carriers, switches, sockets, plugs, etc.
Air	Although electricity can arc through air, this is rare and air is considered a good insulator	The barrier between uninsulated HV overhead cables and anyone below. The insulating layer on a variable capacitor

*No longer used due to health and safety issues
**Still used as overhead HV cable insulators but less and less

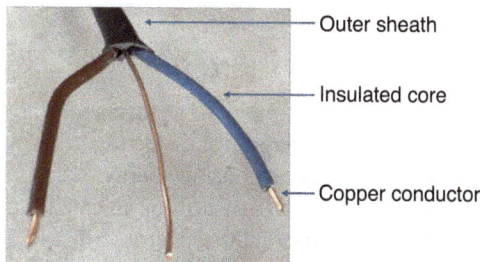

Figure 3.4   Twin-and-protective-conductor PVC-insulated PVC-sheathed wiring cable

damage. The conductors are made of copper, which is softened by annealing to make it flexible. The flexibility is sometimes further improved by using stranded instead of solid conductors.

Semiconductors

Semiconductors have electrical properties that lie between those of conductors and insulators. They occupy a very important place in such devices as LED lamps, rectifiers and transistors, which will be considered in Chapter 18.

3.5   Electrical energy, work and power

Before we can go on to consider the electrical force that results in electron drift in a conductor, we must look at the units used for measuring work and power. A deeper consideration of electrical energy, work and power is given in Chapter 5.

Energy and work (symbol, w; measured in joules, J)

Energy (w) and work are interchangeable, because energy is used to do work. Both are measured in terms of force and distance. The SI unit for:

- Force (F) is the newton (N)
- Distance or length (l) is the metre (m).

If a force moves through a distance, work is done and energy is used. Energy is measured in joules (J):

Energy used = Work done
= Force needed for movement (N) × Distance moved (m)
$$w = Fd$$

Example 3.4
A force of 2000 N is required to lift a machine. How much energy is required to lift the machine through 3 m?

Work = Force × Distance $w = Fd$ $w = 3 \times 2000 = 6000$ J

Power (symbol, P; measured in watts, W)

The energy or work calculated in Example 3.4 is mechanical, but work can also be electrical. For example, consumers pay for the electrical energy that they use.

Power is the rate of doing work or of using energy. For instance, an electrician can cut a hole in a steel plate using a hand drill or an electric drill. With both, the effective work done will be the same, but the electric drill will cut through the hole more quickly because its power is greater.

It follows that Power = $\dfrac{\text{Energy}}{\text{Time period used}}$ $P = \dfrac{w}{t}$

The SI unit of power is the watt (W), which is a rate of doing work of one joule per second.

$$\text{watts} = \frac{\text{joules}}{\text{seconds}}$$

Similarly, if we know how much power is being used and the time for which it is used, we can find the total energy required:

Work or energy $=$ power $\times$ time $w = Pt$ or joules $=$ watts $\times$ second

This is why electricity is paid for as kWh units. It is the amount of energy used over a period of time.

Example 3.5

An installation dissipates 2 kW of electrical energy each hour over a three hour period. How many energy units of electricity have been used?

Units of electricity are kWh.

$w = Pt$ we can use hours as our time in this case. $w = 2000 \times 3$

$= 6000$ J or 6 kWh

3.6   Electromotive force and potential difference

Electromotive force (electromotive force (EMF) symbol E; measured in volts V)

When an electric current flows, energy is dissipated. Since energy cannot be created, it must be provided by the device used for circulating the current. This device may be:

- chemical, such as a battery;
- mechanical, such as a generator.

Many years ago, electricity was thought to be a fluid which circulated as the result of a force, and thus the term electromotive force (EMF), symbol E, came into use. EMF is measured in terms of the number of joules of work (w) necessary to move one coulomb of electrical charge (Q) around the circuit and, because of this, has the unit joules/coulomb. However, this unit is generally referred to as the volt (symbol V), so that

$$E = \frac{w}{Q} \quad w = EQ \quad Q = \frac{w}{E}$$

If 1 J or energy results in 1 C of charge $E = \frac{I}{V} \quad E = IV$

Example 3.6
A battery with an EMF of 6 V gives a current of 5 A around a circuit for 5 min.
How much energy is provided in this time?

Total charge transferred $Q = It = 5 \times (5 \times 60) = 1500 \text{ C}$

Total energy supplied $=$ volts $\times$ coulombs $w = EQ = 6 \times 1500 = 9000 \text{ J}$

In Example 3.6, each coulomb of electricity contained 6 J of potential energy
on leaving the battery. This energy was dissipated on the journey around the circuit
so that the same charge would possess no energy on its return to the battery.

Potential difference (symbol, V; measured in volts, V)

In simple terms, potential difference is the result of the difference in charge
between the positive and negative terminals of the supply. A circuit fed by a battery
is the easiest type of circuit to illustrate this point. The more powerful the battery,
the greater the difference in levels of positive and negative charge on its positive
and negative terminals. As a result, more EMF will be produced.

Think of it as a waterfall. The higher the waterfall, the greater the difference
between the river feeding into the waterfall and the river running away from it at
the bottom, and the more powerful the flow of water.

In terms of energy, the amount expended by 1 C in its passage between any
two points in a circuit is known as the potential difference (pd) between those
points, and is measured in joules/coulomb or volts.

The international symbol for supply voltage or voltage drop is U, the symbol
for the unit of voltage (the volt) is V. A convenient definition of the volt is therefore
that it is equal to the difference in potential between two points if 1 J of energy is
required to transfer 1 C of electricity between them.

Example 3.7
How much electrical energy is converted into heat each minute by an electric heater
that takes 13 A from a 230 V supply?

Energy given up by each coulomb (J/C) $= 230$ J also known as 230 V.
Quantity of energy flow per minute $= It = 13 \times 60 = 780$C
Energy converted in 1 min $=$ (joules/coulomb)
$\times$ coulombs or volts $\times$ coulombs
$w = w/Q \times Q$ or $w = V \times Q$
$= 230 \times 780$ joules
$= 179{,}400$ J or 179.4 kJ

3.7 Current (symbol, *I;* measurement, ampere A – more commonly called amp)

We have already discussed the nature of electrical current in Sections 3.1 and 3.2. Current is essentially the flow of electricity along a conductor. It is that flow of current that causes an electrical device to operate. Current is proportional to voltage and inversely proportional to resistance; in other words, the higher the resistance the lower the current. This can be seen from the formula:

$$I = \frac{U}{R} \quad I = \frac{50}{25} \quad I = 2\,A$$

$$I = \frac{50}{40} \quad I = 1.25\,A$$

When electrical energy is used, power is dissipated and voltage and current are inversely proportional. The higher the voltage, the lower the current. This can be seen from the formula:

$$I = \frac{P}{U} \quad I = \frac{300}{50} \quad I = 6\,A$$

$$I = \frac{300}{100} \quad I = 3\,A$$

Current, rather than voltage, is the cause of injury due to electric shock. A mere 40 mA is considered to be fatal. Most protective devices installed into a circuit or installation are designed to operate on overcurrent faults, that is, a sudden and catastrophic rise in current due to a short circuit or earth fault (see Chapter 14).

3.8 Resistance: Ohm's Law (symbol, *R*; measured in ohms, Ω)

If a metallic conductor is kept at a constant temperature, you can apply the ratio:

$$\frac{\text{Potential difference across conductor (volts)}}{\text{Resulting current in conductor (amperes)}}$$

This ratio is known as the resistance (symbol *R*) of the conductor. This important relationship was first verified by Dr G. S. Ohm and is often referred to as Ohm's Law. The unit of resistance is the ohm (Greek symbol Ω, 'omega').

A conductor has a resistance of 1 Ω if the potential difference across its ends is 1 V when it carries a current of 1 A.

Resistance is an inherent property of a material that affects the amount of current that can flow through it. An insulator has an extremely high resistance, so high, in fact, virtually no current can pass through it at all. A conductor has a very low resistance, but it *does* have resistance. Copper, for example, has a resistance, or resistivity, of 1.72×10^{-8} Ω/m.

Levels of resistance are dependent on conductor:

- Temperature – increased temperature = higher resistance
- Length – increased length = higher resistance
- Cross-sectional area – increased cross-sectional area = lower resistance

Every electrical load, be it a phone charger, luminaire or large machine in a factory, has a resistance that limits current flow through the circuit. If there was no load, the current would be so high that it would cause an explosive eruption of heat and melt the conductor – an occurrence better known as a short circuit. Some circuits include components which have a set resistance, intended to limit the current flow to a certain level. These components are called resistors.

The relationship expressed by Ohm's Law, which is of fundamental importance in electrical engineering, can be written as the formula:

$$R = \frac{U}{I}$$

The subject of the formula can be transposed to give

$$I = \frac{U}{R} \quad U = IR$$

Example 3.8
An electrical heater used on a 230 V supply carries a current of 12 A. What is its resistance?

$$R = \frac{U}{I} = \frac{230}{12} = 19.2 \; \Omega$$

Example 3.9
The insulation resistance between two cables is two million ohms (2 M Ω). What leakage current will flow if a PD of 400 V exists between them?

$$I = \frac{U}{R} = \frac{400}{2,000,000} = 0.0002 \text{ A or } 0.2 \text{ mA}$$

Example 3.10
What PD exists across an earth-continuity conductor of resistance 1.2 Ω when a current of 25 A flows through it?

$$U = IR = 25 \times 1.2 = 30 \text{ V}$$

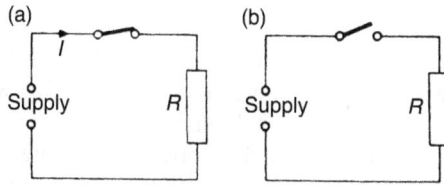

Figure 3.5 EMF in a circuit: (a) closed circuit and (b) open circuit

3.9 Electrical circuit

A closed circuit is necessary for electricity to flow. In practice, the circuit will consist of an item of electrical equipment connected to the source of EMF by means of cables. Figure 3.5 shows a simple circuit consisting of a source of EMF and a resistor.

A switch (shown closed in Figure 3.5(a)) is included as a means of breaking and making the circuit. With the switch closed, the circuit is completed, enabling the supply EMF to provide a current. The current value will depend on the voltage of the supply and the resistance of the circuit.

- If the load has a high resistance, the current will be small.
- If the load resistance is low, the current will be greater.

If the switch is opened, as in Figure 3.5(b), the gap between the opened contacts introduces an almost infinitely high resistance into the circuit. The current will fall to zero. We say that opening the switch has 'broken' the circuit.

Although there are some devices, including electronic circuits and fluorescent luminaires, which do not offer a complete metallic path for current, the majority of circuits are made up entirely of conductors. If the conducting path is interrupted, the current ceases to flow.

Switching circuits where high voltages are present may not break the circuit. The current might continue to flow through the air between the contacts. The air carrying the current glows brightly and gives off heat. This is called an arc. The majority of switches produce an arc when opened, but in most cases the arc disappears within a fraction of a second. The circuit is broken before the heat from the arc can damage the switch and its surroundings.

A loose connection can also cause an arc. This results in a hot spot in the circuit and could cause a fire if left unrepaired. An Arc Fault Detection Device (AFDD) can be introduced into a circuit to detect this type of fault before too much damage has been caused (see Chapter 14).

3.10 Ammeters and voltmeters

Although the presence of an electric current may produce effects that can be detected by the human senses, such effects are seldom useful as an indication of the value of the current. For instance, when a lamp glows, it is clearly carrying current, although we are unlikely to be able to state its value.

Ammeters

Instruments for direct measurement of electric current are called ammeters. These instruments have low resistance and are connected in series within the circuit so that the current to be measured passes through them.

Figure 3.6 shows correct (a) and incorrect (b) ammeter connections. The ammeter will be damaged if incorrectly connected. Figure 3.6 makes it clear that the circuit symbol for an ammeter is a circle containing the letter A.

Because it is often impractical to open a working circuit to insert an ammeter, the clamp-on ammeter (Figure 3.7) is often used to measure current.

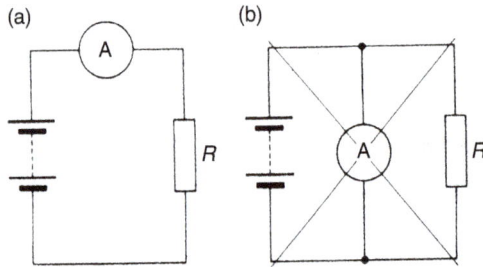

Figure 3.6   Correct (a) and incorrect (b) connection of ammeter

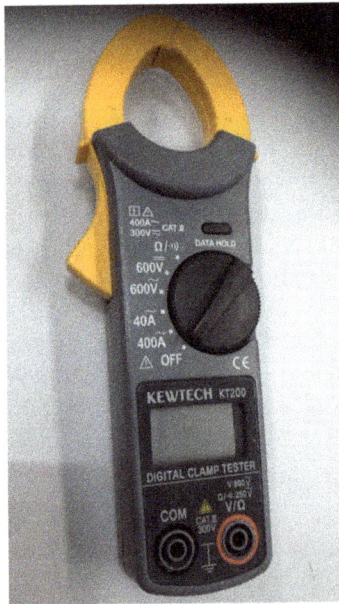

Figure 3.7   Clamp-on ammeter (by kind permission of Oaklands College, St Albans)

(a)   (b)

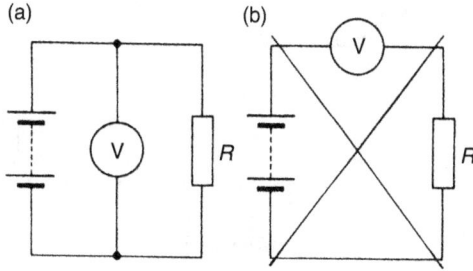

Figure 3.8   Correct (a) and incorrect (b) connection of voltmeter

Voltmeters

A voltmeter is used to indicate the potential difference between its two connections. The voltmeter must be connected in parallel across the device or circuit whose PD is to be measured. Figure 3.8 gives correct (a) and incorrect (b) voltmeter connections and shows the voltmeter symbol as a circle containing the letter V. If connected incorrectly, the voltmeter is unlikely to be damaged but since it has high resistance, it will prevent the correct functioning of the circuit.

3.11   Series circuits

When a number of resistors are connected together end-to-end so that there is only one path for current through them, they are said to be connected in series. In a series circuit:

- The current is the same wherever it is measured.
- The supply voltage is divided up across the resistances, for example:

$$U_{supply} = U \text{ across } R_1 + U \text{ across } R_2 + U \text{ across } R_3, \text{ etc.}$$

Loads, such as luminaires in general installation circuits, are seldom wired in series, because a break in any part of the circuit, a blown lamp for example, would cause the whole circuit to fail.

Each individual light, however, is wired in series with the supply, as are electrical appliances. Because the total current will depend on the resistance of the circuit as well as on the voltage applied to it, it is important to be able to calculate the resistance of the complete circuit if we know the values of its individual resistors.

Figure 3.9 shows three resistors of values R_1, R_2 and R_3, respectively, connected in series across a supply of U volts. Let us assume that the resulting current is I amperes. If the total circuit resistance is R ohms,

$$R = \frac{U}{I}$$

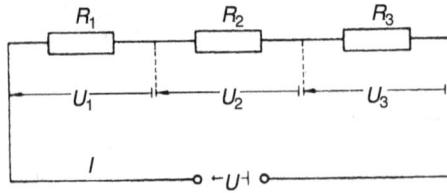

Figure 3.9   Resistors connected in series

Now let the PD across each of the three resistors be U_1, U_2 and U_3 volts, respectively. Then:

$$U_1 = IR_1, U_2 = IR_2 \text{ and } U_3 = IR_3$$

but

$$\begin{aligned} U &= U_1 + U_2 + U_3 \\ &= IR_1 + IR_2 + IR_3 \\ &= I \times (R_1 + R_2 + R_3) \end{aligned}$$
$$\frac{U}{I} = R_1 + R_2 + R_3$$

Thus, the total resistance of any number of transistors connected together in series can be found by adding the values of the individual resistors, which must all be expressed in the same unit.

Example 3.11
Resistors of 50 Ω and 70 Ω are connected in series to a 230 V supply. Calculate

(a) the total resistance of the circuit
(b) the current
(c) the PD across each resistor.

(a) $R = R_1 + R_2 = 50 + 70 = 120$

(b) $I = \dfrac{U}{R} = \dfrac{230}{120} = 1.92$ A

(c) $U_1 = IR_1 = 1.92 \times 50 = 96$ V

$\quad U_2 = IR_2 = 1.92 \times 70 = 134$ V

Note: Supply voltage $U = U_1 + U_2 = 96 + 134$ volts $= 230$ V
    Note that, for a series circuit:

1. the same current flows in all resistors
2. the PD across each resistor is proportional to its resistance
3. the sum of the PDs across individual resistors is equal to the supply voltage.

Example 3.12

An electric heater consists of an element of resistance 22.8 Ω and is fed from a 230 V supply by a two-core cable of unknown resistance. If the current is 10 A, calculate the total resistance of the cable.

Figure 3.10(a) shows the heater connected to the supply through a two-conductor cable. Figure 3.10(b) is an equivalent circuit in which conductors are assumed to have no resistance, the actual resistance of the conductors being replaced by the resistance R. There are two methods of solution.

Method 1

$$\text{Total resistance} = \frac{U}{I} = \frac{230}{10} = 23 \ \Omega$$

But

$$\text{Total resistance} = \text{Element resistance} + \text{Conductor resistance}$$

Therefore,

$$\text{Conductor resistance} = \text{Total resistance} - \text{Element resistance}$$
$$\text{Conductor resistance} = 23 - 22.8 = 0.2 \ \Omega$$

Method 2

$$\text{PD across element} = \text{Current} \times \text{Element resistance} = 10 \times 22.8 = 228 \ \text{V}$$
$$\text{Supply voltage} = \text{PD across element} + \text{PD across conductors}$$

Therefore,

$$\text{PD across conductors} = \text{Supply voltage} - \text{PD across element}$$

$$= 230 - 228 \ \text{V} = 2 \ \text{V}$$

$$\text{Conductor resistance} = \frac{\text{conductor PD}}{\text{current}} = \frac{2}{10} = 0.2 \ \Omega$$

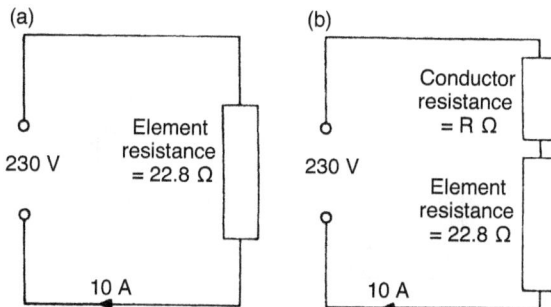

Figure 3.10 Circuit diagrams for Example 3.12

3.12   Parallel circuits

When each one of a number of resistors is connected between the same two points, they are said to be connected in parallel. Another way to look at a parallel circuit is that each load is supplied by its own individual 'sub-circuit'. These 'sub-circuits' are then collected together and connected to the same point of supply. For parallel circuits:

- The voltage is the same at every point in the circuit.
- The total current is divided between the resistors.

Since all the resistors are connected across the same two points, the PD across each one is the same. Figure 3.11 shows resistors of values R_1, R_2 and R_3, respectively, and the total current is I.

The total current divides itself among the resistors, so that:

$$I = I_1 + I_2 + I_3$$

But, from Ohm's law,

$$I_1 = \frac{U}{R_1} I_2 = \frac{U}{R_2} I_3 = \frac{U}{R_3}$$

$$I = \frac{U}{R_1} + \frac{U}{R_2} + \frac{U}{R_3} = U \times \frac{1}{R_1} + \frac{1}{R_2} + \frac{1}{R_3}$$

$$\frac{I}{U} = \frac{U}{R_1} + \frac{U}{R_2} + \frac{U}{R_3}$$

If the equivalent resistance of the parallel circuit is R, $R = \frac{U}{I}$
Therefore,

$$\frac{U}{I} = \frac{1}{R} = \frac{1}{R_1} + \frac{1}{R_2} + \frac{1}{R_3}$$

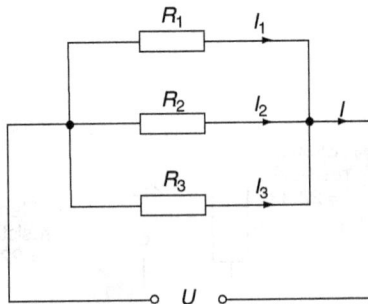

Figure 3.11   Resistors connected in parallel

The value below is called the reciprocal of R. We can sum up the expression by saying that the reciprocal of the equivalent resistance of a parallel circuit is equal to the sum of the reciprocals of the resistances of the individual resistors.

$$\frac{1}{R}$$

Note that, for a parallel circuit:

1. the same PD occurs across all resistors
2. the current in each resistor is inversely proportional to its resistance
3. the sum of the currents in the individual resistors is equal to the supply current.

As for the series circuit, the resistances must all be expressed in the same unit before using them in the formula.

Example 3.13
Calculate the equivalent resistance of three parallel-connected resistors of 6 Ω, 30 Ω and 10 Ω, respectively.

$$\frac{1}{R} = \frac{1}{R_1} + \frac{1}{R_2} + \frac{1}{R_3}$$

$$\frac{1}{R} = \frac{1}{6} + \frac{1}{30} + \frac{1}{10}$$

Common denominator = 30

$$\frac{5+1+3}{30} = \frac{9}{30} \quad R = \frac{30}{9} \quad R = 3.33\ \Omega$$

It can be seen from this result that the equivalent resistance of any group of parallel-connected resistors is lower than that of the lowest-valued resistor in the group.

If a number of equal-value resistors are connected in parallel, the equivalent resistance will be the value of one resistor divided by the number of resistors. For example, five 10 Ω resistors in parallel have an equivalent resistance of $\frac{10}{5} = 2\ \Omega$.

It has been simple to find the lowest common denominator in Example 3.13 because all three resistor values divide exactly into the maximum value of 30 Ω. If the resistors had less convenient values with no obvious common denominator, the process would have been awkward. Fortunately modern scientific calculators include a reciprocation function, marked X^{-1}. Any number with an index of minus one is reciprocated (see Chapter 1).

E.g. $20^{-1} = \dfrac{1}{20}$

Use the button after entering each value. Remember, the answer is also a reciprocated number, so use the X^{-1} button to correct this value.

Example 3.14
Three resistors are connected in parallel. Their values are 13 Ω, 18 Ω and 22 Ω respectively. To calculate the equivalent resistance using the reciprocate button on a calculator.
Enter:

$$13(X^{-1}) + 18(X^{-1}) + 22(X^{-1}) = 0.18(X^{-1}) = 5.62 \ \Omega$$

Example 3.15
Resistors of 16 Ω, 24 Ω and 48 Ω, respectively, are connected in parallel to a 230 V supply. Calculate the total current.
There are two ways of solving the problem.

Method 1
Find the equivalent resistance and use it with the supply voltage to find the total current.

$$\frac{1}{R} = \frac{1}{16} + \frac{1}{24} + \frac{1}{48} = \frac{3+2+1}{48} = \frac{6}{48}$$

$$R = \frac{48}{6} = 8 \ \Omega$$

$$I = \frac{U}{R} = \frac{230}{8} = 28.75 \ \text{A}$$

Method 2
Find the current in each resistor. Add these currents to give the total current.

$$\text{Current in 16 } \Omega \text{ resistor} = \frac{U}{R} = \frac{230}{16} = 14.38 \ \text{A}$$

$$\text{Current in 24 } \Omega \text{ resistor} = \frac{U}{R} = \frac{230}{24} = 9.58 \ \text{A}$$

$$\text{Current in 48 } \Omega \text{ resistor} = \frac{U}{R} = \frac{230}{48} = 4.79 \ \text{A}$$

$$\text{Total current} = 14.38 + 9.58 + 4.79 = 28.75 \ \text{A}$$

There are many applications of parallel circuits. The elements of a two-bar fire are connected in parallel, and the heat output is varied by switching one bar on or off as required. The two circuits in a cooker grill can be connected in three ways to give three-heat control, as indicated in the following example.

Example 3.16

The grill of an electric oven has two identical elements, each of resistance 48 Ω, which are connected in parallel for 'high' heat and in series for 'low' heat. One element only is used for 'medium' heat. Calculate

(a) The current drawn from a 230 V supply for each switch position.
(b) The power dissipated by each setting

Low
The elements are in series (Figure 3.12(a))

Total resistance = 48 + 48 ohms = 96 Ω

Therefore,

$$I = \frac{U}{R} = \frac{230}{96} = 2.4 \text{ A} \quad P = UI \quad P = 230 \times 2.4 = 552 \text{ W}$$

Medium
One element only in use (Figure 3.12(b)). Therefore,

$$I = \frac{U}{R} = \frac{230}{48} = 4.79 \text{ A} \quad P = UI \text{ P} = 230 \times 4.79 = 1.1 \text{ kW}$$

High
The elements are in parallel (Figure 3.12(c)). They are of equal resistance so it can be calculated as:

$$\frac{\text{Resistance of one element}}{\text{Number of elements}} = \frac{48}{2} = 24 \text{ A} \quad P = UI \quad P = 230 \times 24 = 5.52 \text{ kW}$$

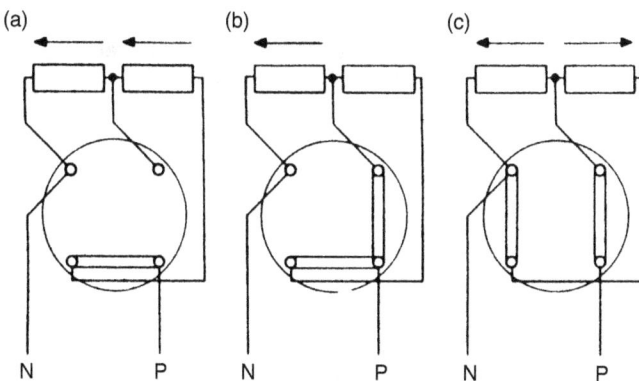

*Figure 3.12   Three-heat switching circuits for Example 1.16: (a) low;
(b) medium; and (c) high*

3.13   Series–parallel circuits

A series–parallel circuit is one which is made up of series and parallel parts in combination. The possible number of combinations is endless, but all these circuits can be solved by **simplification**. A number of resistors are seen to be in series or in parallel. These can be replaced by one resistor value which has the same effect on the circuit. This principle is explained in Example 3.16.

> **KEY TERM: Simplification – Carrying out calculations to make a mathematical problem more compact and manageable.**
>
> For example, $\dfrac{15}{25}$ can be simplified because both numbers are divisible
>
> by 5 $\dfrac{15}{25}$ becomes $\dfrac{3}{5}$

Example 3.17
Two banks of resistors are connected in series. The first bank consists of two resistors of 10 Ω and 40 Ω in parallel, and the second consists of three resistors, each of 12 Ω, connected in parallel. What is the resistance of the combination, and what current will be taken from a 12 V supply to which it is connected?

The circuit diagram is shown in Figure 3.13(a). To solve, we must look for groups of resistors connected in series or in parallel. The first bank consists of two resistors in parallel, so we must find the resistance of this combination.

$$\frac{1}{R} = \frac{1}{10} + \frac{1}{40} + \frac{4+1}{40} = \frac{5}{40} \quad R = \frac{40}{5} = 8\,\Omega$$

The first group of resistors can be replaced by a single 8 Ω resistor. The second bank consists of three 12 Ω resistors in parallel. Its resistance can be found using reciprocation as above but also by dividing the value of one resistor

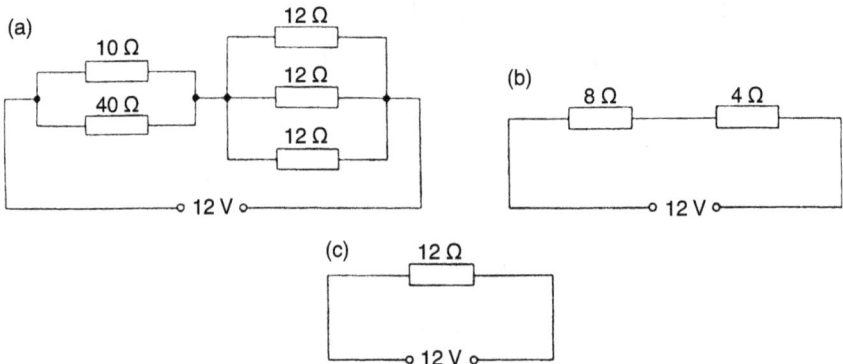

Figure 3.13   Circuit diagrams for Example 3.17

by the number of resistors:

$$R = \frac{12}{3} = 4\,\Omega$$

The second group can this be replaced by a single 4 Ω resistor. Figure 3.13(b) shows a simple series circuit which is the equivalent of that in Figure 3.13(a). These two circuits are not identical, but they have the same resistance and will take the same current when connected to a supply.

These two resistors can now be combined to a single equivalent resistor:

$$R = 8 + 4 = 12\,\Omega$$

This is the resistance of the complete circuit. To find the current, Ohm's law is applied:

$$I = \frac{U}{R} = \frac{12}{12} = 1\,\text{A}$$

Example 3.18

Figure 3.14 shows a resistor network connected to a 230 V supply. Ammeters and voltmeters are to be connected to measure the current in each resistor and the PD across each resistor. Redraw the diagram, adding these instruments. Calculate their readings.

Ammeters must be connected so that the current to be measured passes through them. An ammeter must therefore be connected in series with each resistor. Voltmeters must be connected so that the potential difference to be measured is also across their terminals. A voltmeter could thus be connected in parallel with each resistor.

Resistors connected in parallel clearly have the same potential difference across them, so one voltmeter connected across such a parallel group will suffice.

Figure 3.14(a) shows the circuit with ammeters and voltmeters added. It should be noted that voltmeter V_x is connected across

- The 80 Ω resistor
- Ammeter A1
- The 120 Ω resistor
- Ammeter A2

Each of the ammeters has resistance, and hence a potential difference appears across it when carrying current. The voltmeter will read the sum of the PDs across a resistor and an ammeter. In practice, the resistance of an ammeter is so small that in most cases the potential difference across it can be ignored.

To find the readings on the instruments, we must calculate the PD across each resistor, as well as the current through each resistor. The first step is to find the resistance of the whole circuit and hence the total current. It is most important when dealing with circuits of this sort to be quite clear which part of

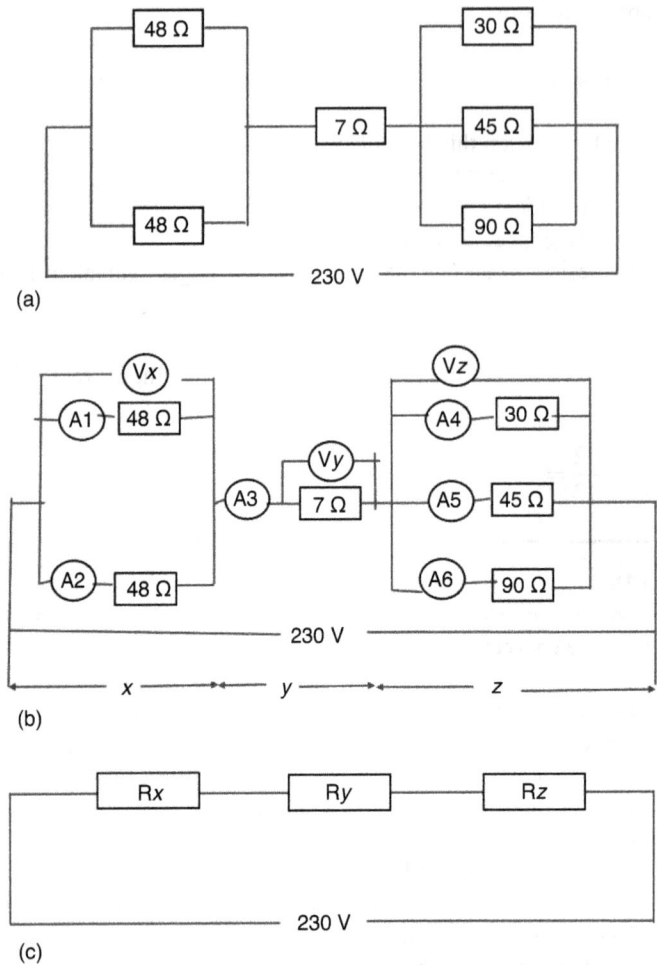

Figure 3.14   Circuits for solution of Example 3.18

the circuit is being considered. To this end, the three-series-connected sections of the circuit have been called *x*, *y* and *z*, respectively (Figure 3.14(a)). As in the previous example, the circuit must now be reduced to the simple series circuit shown in Figure 3.14(b).

STEP 1: Find the equivalent resistance for R_x, R_y and R_z

$$R_x = \frac{1}{48} + \frac{1}{48} = \frac{2}{48} = 2\,\Omega$$

Alternatively, as there are two identical resistances: $\dfrac{48}{2} = 24 \ \Omega$

$R_y = 7 \ \Omega$

$$R_z = \frac{1}{30} + \frac{1}{45} = \frac{1}{90} = \frac{3+2+1}{90} = \frac{6}{90} = 15 \ \Omega$$

Total circuit resistance $= R_x + R_y + R_z = 24 + 7 + 15 = 46 \ \Omega$

STEP 2: Calculate supply current

Current from supply, $I = \dfrac{U}{R}$ $I = \dfrac{230}{46} = 5 \ \text{A}$

STEP 3: Calculate volt drop across each section

When this current flows in the circuit of Figure 3.14(b), the voltage drop across R_x is:

$$U_x = IR_x = 5 \times 24 \text{ volts } = 120 \text{ V}$$

Since voltmeter $V\times$ is connected across a group of resistors having the same effect as R_x:

$V\times$ will read 120 V.
Similarly $U_y = IR_y = 5 \times 7 = 35$ V
Voltmeter V_y will read 35 V.
Similarly $U_z = IR_z = 5 \times 15 = 75$ V
Voltmeter V_z will read 75 V. Note that as pointed out in Section 3.10, for a series circuit, the sum of the PDs across the individual resistors is equal to the supply voltage. To check this:

$$120 \text{ V } + 35 \text{ V } + 75 \text{ V } = 230 \text{ V}$$

STEP 4: Calculate current in each resistor

We can now calculate the current in each resistor.

Section x

The resistors in section x each have the section PD of 120 V applied to them.

$$I_1 = \frac{U_x}{R_1}$$

$$= \frac{120}{48} = 2.5 \text{ A}$$

Ammeter A1 reads 5 A.

$$I_2 = \frac{U_x}{R_2}$$

$$= \frac{120}{48} = 2.5 \text{ A}$$

Ammeter A2 reads 2.5 A.

Note that as pointed out in Section 3.11, for a parallel circuit the sum of the currents in individual resistors is equal to the supply current. To check this:

$$2.5 \text{ A} + 2.5 \text{ A} = 3 \text{ A}$$

Section y

The 7 Ω resistor in section y has 35 V applied to it.

$$I_3 = \frac{U_y}{R_3}$$

$$\frac{35}{7} \text{ amperes} = 5 \text{ A}$$

Ammeter A3 reads 5 A.

Inspection of the circuit would have confirmed this current without calculation.

Since the 17 Ω resistor has no other resistor in parallel with it, it must carry the whole of the circuit current.

Section z

The resistors in section z, each have 75 V applied to them. Thus:

$$I_4 = \frac{U_z}{R_4}$$

$$= \frac{70}{30} = 2.5 \text{ A}$$

and ammeter A4 reads 2.5 A;

$$I_5 = \frac{U_z}{R_5}$$

$$= \frac{75}{45} \text{ amperes} = 1.67 \text{ A}$$

and ammeter A5 reads 1.67 A; and

$$I_6 = \frac{U_z}{R_6}$$

$$= \frac{75}{90} \text{ amperes} = 0.83 \text{ A}$$

and ammeter A6 reads 0.83 A. To check:

$$2.5 \text{ A} + 1.67 \text{ A} + 0.83 \text{ A} = 5 \text{ A}$$

Example 3.19

The circuit shown in Figure 3.15 takes a current of 6 A from the 50 V supply. Calculate the value of resistor R_4.

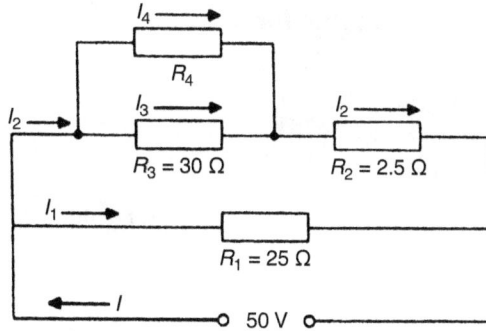

Figure 3.15   Diagram for Example 3.19

There are several ways of tackling this problem. One of the simplest demonstrates a different approach from those possible in the previous worked examples. First, mark the currents on the diagram (this has already been done in Figure 3.15).

Find I_1

R_1 has the 50 V supply directly across it, so:

$$I_1 = \frac{U}{R_1} = \frac{50}{25} = 2 \text{ A}$$

Since the total current is 6 A, it follows that the current I_2 in the upper part of the circuit is

$$6 \text{ A} - 2 \text{ A} = 4 \text{ A.}$$

Find the PD (U_2) across R_2

$$U_2 = I_2 R_2 = 4 \times 2.5 \text{ V} = 10 \text{ V.}$$

The parallel combination of R_3 and R_4 is in series with R_2 across the 50 V supply.

Find I_3

The PD across R_2 is 10 V.

The PD across both R_3 and R_4 must be 50 V $-$ 10 V $= 40$ V.

$$I_3 = \frac{U_3}{R_3} = \frac{40}{30} = 1.33 \text{ A}$$

Find I_4

$$I_4 = I_2 - I_3 = 4 - 1.33 = 2.67 \text{ A}$$

Applying Ohm's law to R_4:

$$R_4 = \frac{U_4}{R_4} = \frac{40}{2.67} = 15 \text{ }\Omega$$

3.14   Summary of formulae for Chapter 3

Electrical charge

$$Q = It \quad I = \frac{Q}{t} \quad t = \frac{Q}{I}$$

where Q = electric charge (C); I = current (A); and t = duration of current (s).

Work

$$W = Fd \quad d = \frac{W}{F} \quad F = \frac{W}{d}$$

where W = energy used or work done (J); F = force applied (N); and d = distance moved (m).

Power

$$P = \frac{w}{t} \quad w = Pt \quad t = \frac{w}{P}$$

$$P = UI \quad U = \frac{P}{I} \quad I = \frac{U}{I}$$

where P = power or rate of doing work (W); w = work (J); t = time (s); U = applied voltage (V); and I = current (A).

Potential difference and charge

$$w = UQ \quad U = \frac{w}{U} \quad Q = \frac{w}{U}$$

where U = potential difference (V); Q = electrical charge (C); and w = work (J).

Resistors in series

$$R = R_1 + R_2 + R_3 + \ldots$$

Resistors in parallel

$$\frac{1}{R} = \frac{1}{R_1} + \frac{1}{R_2} + \frac{1}{R_3} \text{ or } R^{-1}$$

$$= R_1^{-1} + R_2^{-1} + R_3^{-1} \left(X^{-1} \text{ button on a scientific calculator} \right)$$

Ohm's Law

$$I = \frac{U}{R} \quad R = \frac{U}{I} \quad U = IR$$

where U = applied voltage (V); I = current (A); R = circuit resistance (Ω).

3.15 Exercises

1. A current of 10 A flows for 2 min. What charge is transferred?
2. For how long must a current of 4 mA flow so as to transfer a charge of 24 C?
3. What current must flow if 100 C is to be transferred in 8 s?
4. Briefly describe one example of each of the chemical, heating and electro-magnetic effects of an electric current.
5. A DC generator has an EMF of 200 V and provides a current of 10 A. How much energy does it provide each minute?
6. A photocell causes a current of 4 μA in its associated circuit, and would take 1000 days to dissipate an energy of 1 mJ. What EMF does it provide?
7. If the total resistance of an earth fault loop is 4 Ω, what current will flow in the event of a phase-to-earth fault from 230 V mains?
8. State Ohm's law in words and with symbols.
9. What is the resistance of an heating element that takes 12.0 A from a 230 V supply?
10. The PD across an earth-continuity conductor of resistance 0.3 Ω is found to be 4.5 V. What current is the conductor carrying?
11. Four resistors of values 5 Ω, 15 Ω, 20 Ω and 40 Ω, respectively, are con-nected in series to a 230 V supply. Calculate the resulting current and the PD across each resistor.
12. A 6 Ω resistor and a resistor of unknown value are connected in series to a 12 V supply, when the PD across the 6 Ω resistor is measured as 9 V. What is the value of the unknown resistor?
13. Calculate the resistance of the element of a soldering iron that takes 0.5 A from 230 V mains when connected to them by cables having a total resistance of 0.2 Ω.
14. Calculate the equivalent resistance of each of the following parallel-connected resistor banks:
    (a) 2 Ω and 6 Ω
    (b) 12 MΩ, 6 MΩ and 36 MΩ
    (c) 100 μΩ, 600 μΩ and 0.0012 Ω

15. Answer the following questions by writing down the missing word or words:
    (a) A fuse protects a circuit against ...... and uses the ...... effect of an electric current.
    (b) The unit of quantity of electricity is called the ....
    (c) Two good conductors of electrical current are ...... and ....
    (d) Two items of electrical equipment that use the electromagnetic effect are ...... and ....
    (e) Two insulating materials used in the electrical industry are ...... and ....
    (f) In an electrical circuit, the electron flow is from the ...... terminal to the ...... terminal.
    (g) A bimetallic strip uses the ...... effect of an electric current.

(h) The electron has a ...... charge.

(i) An EMF of 72 V is applied to a circuit, and a current of 12 A flows. The resistance of the circuit is ....

(j) The effective resistance of two 10 Ω resistors connected in parallel is ....

(k) Quantity of electricity = ... × ....

(l) 0.36 amperes = ... milliamperes.

(m) 3.3 kilovolts = ... volts.

16. Three resistors are connected in parallel across a supply of unknown voltage. Resistor A is of 7.5 Ω and carries a current of 4 A. Resistor B is of 10 Ω, and resistor C is of unknown value but carries a current of 10 A. Calculate the supply voltage, the current in resistor B and the value of resistor C.

17. (a) Show, by separate drawings for 'high', 'medium' and 'low' heat positions, the connections of a series–parallel switch controlling two separate sections of resistance wire forming the elements of a heating appliance. (b) If the two sections of resistance wire are of equal resistance, what is the proportional current flow and heating effect in the 'medium' and 'low' positions relative to the 'high' position.

18. Two resistances of 4 and 12 are connected in parallel with each other. A further resistance of 10 is connected in series with the combination. Calculate the respective direct voltages which should be applied across the whole circuit

   (a) to pass 6 A through the 10 resistance

   (b) to pass 6 A through the 12 resistance.

19. A resistance network is connected as shown in Figure 3.16, and takes a total current of 2.4 A from the 24 V supply. Calculate the value of the resistor R_x.

20. The voltmeter shown in Figure 3.17 reads 39 V. What is the value of the resistor across which the voltmeter is connected?

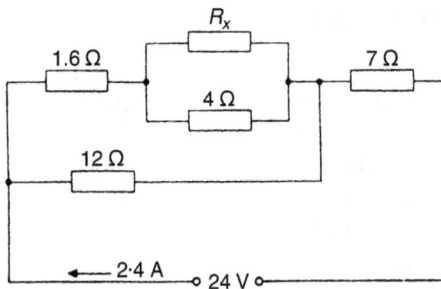

Figure 3.16   Circuit diagram for Exercise 3.19

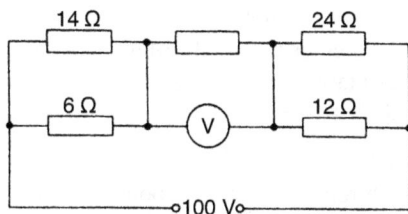

Figure 3.17   Circuit diagram for Exercise 3.20

3.16   Multiple-choice exercises

3M1 The central part of an atom is called the
  (a) neutron
  (b) nucleus
  (c) proton
  (d) electron

3M2 An electric current is considered to consist of
  (a) a drift of electrons from negative to positive
  (b) a movement of positive ions from positive to negative
  (c) an electric fluid travelling from positive to negative
  (d) a drift of electrons from positive to negative

3M3 The unit for the quantity of electricity is called the
  (a) ampere
  (b) watt
  (c) joule
  (d) coulomb

3M4 The heating effect of an electrical current can be used in
  (a) the fuel cell
  (b) the transformer
  (c) an ammeter
  (d) an electric kettle

3M5 The element nickel is
  (a) a soft metal
  (b) an electrical conductor
  (c) a kind of money
  (d) an electrical insulator

3M6 The symbol for energy and work is
  (a) J
  (b) P
  (c) U
  (d) w

3M7   The symbol U is used to denote
    (a) electromotive force or potential difference
    (b) a material unsuitable for electrical use
    (c) electrical resistivity
    (d) a voltmeter

3M8   The formula which relates resistance, potential difference and current is
    (a) $I = UR$
    (b) $R = IU$
    (c) $R = UI$
    (d) $R_T = R_1 + R_2$

3M9   If a current of 26 mA flows in a circuit when a 24 V supply is connected to it, its resistance is
    (a) 923 Ω
    (b) 108 Ω
    (c) 9.23 kΩ
    (d) 92.3 Ω

3M10   To read a current correctly an ammeter must be connected
    (a) so that no current passes through it
    (b) directly across the supply
    (c) in parallel with the circuit concerned
    (d) in series with the circuit concerned

3M11   Two resistors connected in series have a combined resistance of 47 kΩ. If one has a value of 15 k the resistance of the other is
    (a) 62 kΩ
    (b) 32 kΩ
    (c) 22 kΩ
    (d) 3.13 kΩ

3M12   The formula to find the combined resistance of two resistors R_1 and R_2 connected in parallel is
    (a) $R_T = \dfrac{R_1 \times R_2}{R_1 + R_2}$
    (b) $R_T = R_1 + R_2$
    (c) $R_T = \dfrac{R_1 + R_2}{R_1 \times R_2}$
    (d) $R_T = \dfrac{R_1}{R_2}$

Chapter 4

Resistance and resistors

4.1   Introduction

Even conductors have the property of resistance. Usually, this resistance will be small in comparison with that of any other circuit components, and may be ignored for the purposes of many basic circuit calculations. The conductor resistance will be present, however, and as indicated in Section 4.7, the **volt drop** in resistive conductors must be taken into account. If conductors without any resistance at all were possible at a reasonable cost, they would be used widely.

The nearest thing to a perfect conductor is the superconductor, used for research such as in the Large Hadron Collider, but these are too expensive and impractical for widespread commercial use. In many cases, superconductivity is achieved through the application of extremely low temperatures to the conductor. As temperature decreases, so does resistance.

> **KEY TERM: Volt drop – The voltage at the furthest end of a circuit will be lower than the supply voltage. This is due to resistance and current demand. It is important that the volt drop is not high enough to affect the performance of any equipment supplied by the circuit.**

Probably the most common circuit component is the resistor. These components are deliberately made to offer resistance to electric current. Resistors can exist in a very wide variety of shapes and sizes, an example of the most common construction being shown in Figure 4.1. The British Standard objective symbol for a fixed resistor is shown in Figure 4.2(a) with the alternative symbol in Figure 4.2(b)

This symbol has been in use since 1966, is still in use and is likely to be used for many years.

In many practical circuits, as well as in laboratory work, it is often necessary to use a variable resistor, or rheostat. This is usually a resistor with a sliding contact, so that the resistance between the fixed input end and the moveable output can be adjusted by the movement of the slider. Some examples of variable resistors are shown in Figure 4.3.

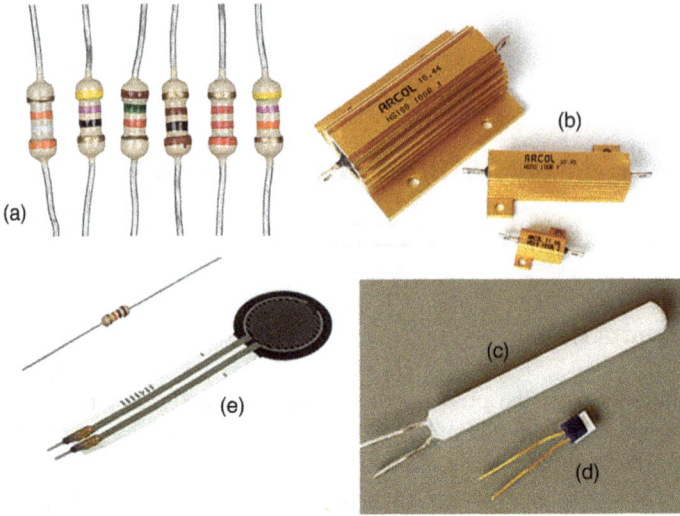

Figure 4.1   *Typical resistors: (a) axial-lead, (b) power, (c) wire-wound, (d) flat film and (e) movement sensing. Image credits: (a) Evan-Amos, Public domain, via Wikimedia Commons; (b–d) Harke, CC BY-SA 3.0, via Wikimedia Commons and (e) Oomlout, CC BY 2.0, via Flickr.*

Figure 4.2   *Circuit symbols for fixed resistor: (a) objective symbol and (b) alternative symbol*

Figure 4.4 shows the objective and alternative general symbols for a variable resistor, and Figure 4.5 shows the symbols that are used when the connection to the slider must be shown.

Many variable resistors can be connected as potential dividers (Figure 4.6) that use the volt drop across resistances to split the supply voltage up into smaller voltages as required. The voltage U_1 that appears at the output terminals will depend on the setting of the slider and on the current taken from the output. Provided that the output current is very small, it follows that:

$$U_1 = \frac{U \times R_1}{R}$$

This method of connection is often used to provide a variable-voltage supply.

(a)

(b)

(c)

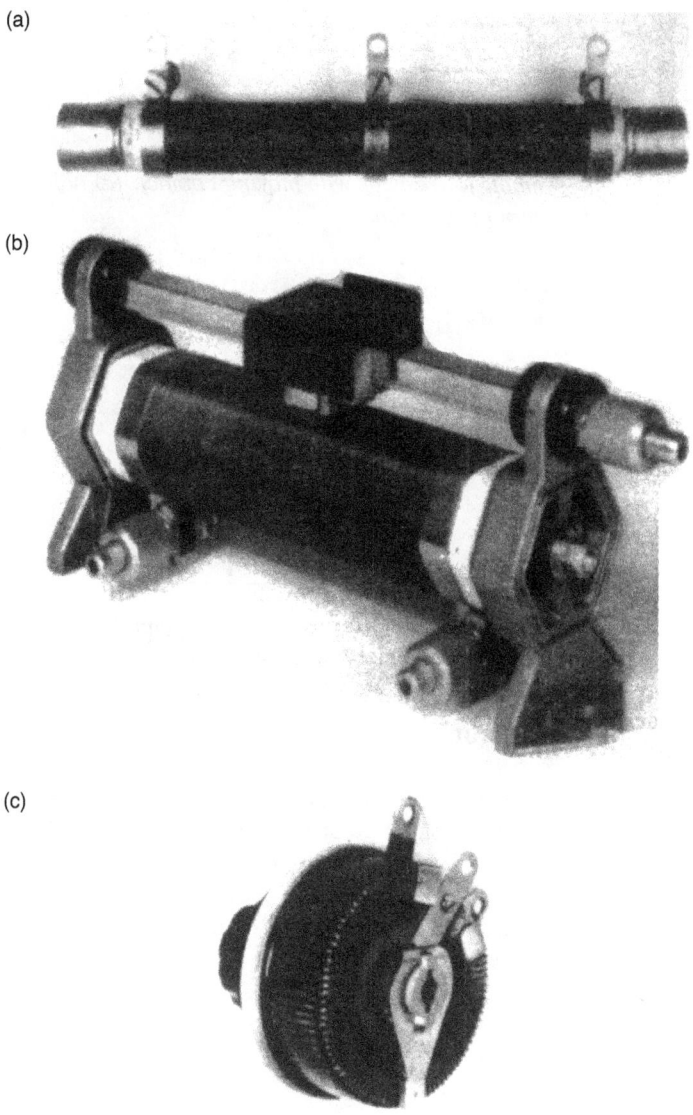

Figure 4.3 *Variable resistors: (a) wire-wound resistor with adjustable tapping;*
(b) wire-wound resistor with continuous variation and (c) wire-wound
toroidal potential divider

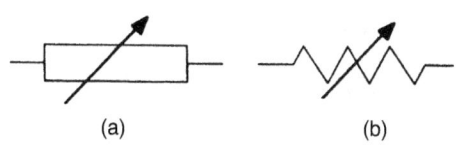

(a) (b)

Figure 4.4 *Circuit symbols for variable resistor: (a) objective symbol and*
(b) alternative symbol

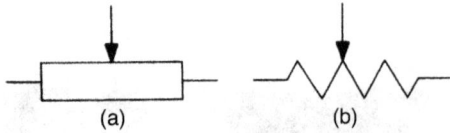

Figure 4.5   *Circuit symbols for resistor with moving contact: (a) objective symbol and (b) alternative symbol*

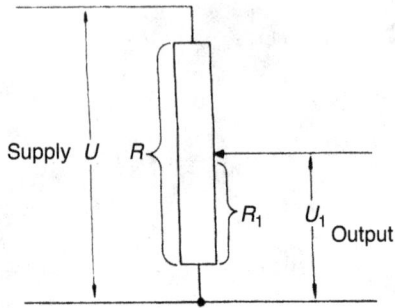

Figure 4.6   *Variable voltage output from potential divider*

4.2   Effect of dimensions on resistance

Let us imagine a cube of conducting material (Figure 4.7(a)) that has a resistance of $r\,\Omega$ between opposite faces. If, say, five of these cubes are joined (Figure 4.7(b)), they form a series resistor chain which has a total of resistance of $5r\,\Omega$.

If four conductors, each made up of five cubes, are placed side by side (Figure 4.7(c)), the result will be one conductor having four times the cross-sectional area of one of the original conductors.

The total resistance R of this composite conductor will be that of four $5r\,\Omega$ resistors in parallel:

$$\frac{1}{R} = \frac{1}{5r} + \frac{1}{5r} + \frac{1}{5r} = \frac{4}{5r}$$

Therefore,

$$R = \frac{5r}{4} \qquad R = 1.25\,\Omega$$

This is the resistance of a conductor five cubes long and of four cubes in cross section. If the conductor had a length l, and a uniform cross-sectional area a, its resistance would be:

$$R = \frac{r \times 1}{a}$$

(a)

$$\equiv \quad -\boxed{r}-$$

(b)

$$\equiv \quad -\boxed{r}-\boxed{r}-\boxed{r}-\boxed{r}-\boxed{r}-$$

$$\equiv \quad -\boxed{5r}-$$

(c)

$$\equiv \quad -\boxed{5r/4}-$$

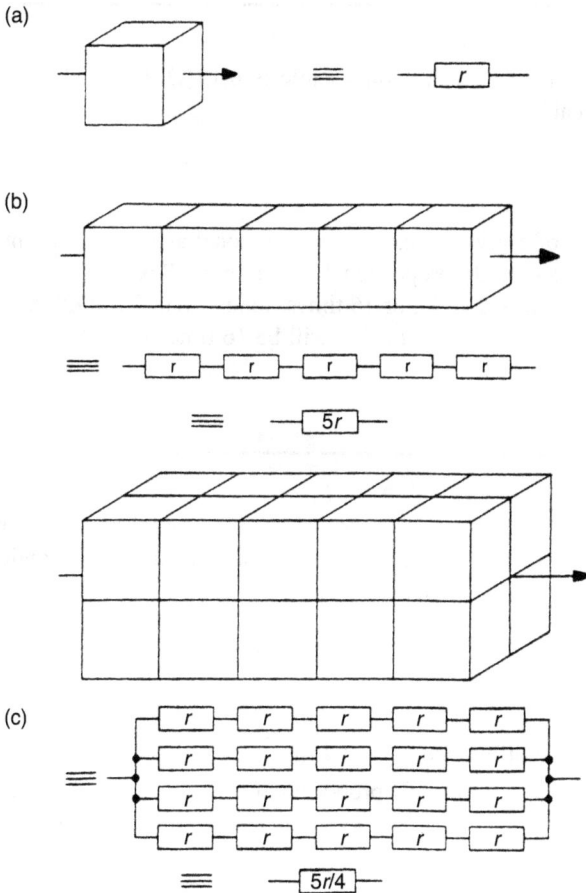

Figure 4.7   Dependence of conductor resistance on dimensions

In this way, the resistance of a conductor can be expressed in terms of its:

- Length
- Cross-sectional area
- Resistance between opposite faces of a cube

Conductors are not often cube-shaped, but the result obtained is true for a cable of any shape if its cross-sectional area is uniform throughout its length.

Resistance is proportional to $\frac{\text{length}}{\text{Area}}$ $R \propto \frac{1}{a}$

Example 4.1
The resistance of 50 m of a certain cable is 0.15 Ω. Calculate the resistance of 800 m of this cable.

$$R \propto \frac{1}{a}$$

- For a cable of a given size, the cross-sectional area a is constant.
- Resistance is directly proportional to length, or $R \propto l$.
- 100 m of cable is 800/50, or 16 times, longer than 50 m of cable.
- The resistance of 800 m of cable will be 16 times that of 50 m of cable.

$$R = 16 \times 0.15 = 2.4 \ \Omega.$$

Example 4.2
If the resistance per 100 m of a cable of cross-sectional area 2.5 mm^2 is 0.06 Ω, what is the resistance of 100 m of a cable made of the same conductor material which has a cross-sectional area of 6 mm^2?

L is constant, so

$$R \propto \frac{1}{a}$$

Calculate size difference $\frac{6}{2.5} = 2.4$ mm^2 is 2.4 times larger than 2.5 mm^2.

Since resistance is inversely proportional to cross-sectional area, the resistance of the 6 mm^2 cable will be 2.4 times smaller, than that of the first cable:

$$R = \frac{0.06}{2.4} = 0.025 \ \Omega$$

Example 4.3
If the resistance (R_1) of 100 m of a 4 mm^2 cable is 0.08 Ω, what is the resistance (R_2) of 700 m of 10 mm^2 cable made of the same conductor material?

$$R \propto \frac{1}{a}$$

Therefore,

$$R_2 = 0.08 \times \frac{700}{100} \times \frac{4}{10} = 0.224 \ \Omega$$

Example 4.4

In the manufacture of aluminium cable, a 10 m length of thick circular rod, which has a resistance of 0.015 Ω, is drawn out until its new diameter is 0.2× its previous measurement. Calculate the length of the drawn cable and its resistance.

By drawing the aluminium bar through a die, its length is increased by reducing its diameter. The volume must, however, be the same before and after the drawing operation, since no metal is lost. The conductor is essentially a cylinder.

V = Volume; V_1 = volume of original rod; V_2 = volume of drawn cable.

$$V_1 = l_1 \times a_1 = l_1 \times \frac{\pi d^2}{4}$$

$$V_2 = l_2 \times a_2 = l_2 \times \frac{\pi d^2}{4}$$

But, in terms of amount of material:

$$V_1 = V_2 \text{ so } l_1 \times \frac{\pi d^2}{4} (V_1) = l_2 \times \frac{\pi d}{4} (V_2)$$

Therefore,

$$l_2 = l_2 \times \frac{\pi d^2}{4} (V_1) \times \frac{4}{\pi d^2} (V_2) = l_1 \times \frac{d^2 (V_1)}{d^2 (V_2)}$$

Since the wire is 0.2× the initial diameter after the drawing process:

$$\frac{d_1}{d_2} = \frac{1}{0.2} = 5$$

Therefore,

$$l_2 = \frac{10}{1^2} \times 5^2 = 10 \times 25 = 250 \text{ m}$$

because

$$R \propto \frac{1}{a} \qquad R_2 = R_1 \times \frac{l_2}{l_1} \times \frac{a_1}{a_2}$$

So $R_1 = 0.0.015 \ \Omega \quad l_1 = 10 \text{ m} \quad l_2 = 250 \text{ m}$

$$a_1 = \frac{\pi d^2}{4} (V_1) = \frac{\pi 1^2}{4} = \frac{\pi}{4}$$

$$a_2 = \frac{\pi d^2}{4} (V_2) = \frac{\pi}{4} \times 0.2^2 = \frac{\pi}{4 \times 25}$$

Therefore,

$$R_2 = 0.015 \times \frac{250}{10} \times \frac{\pi}{4} \times \frac{4 \times 25}{\pi} = 0.0.015 \times 25 = 9.375 \ \Omega$$

4.3   Resistivity

The resistance between opposite faces of a cube of conducting material (given the value r Ω in Section 4.2) is called the resistivity, or sometimes the specific resistance, of the material. For most conductors, this value is very low, and usually measured in microhms (μΩ). For instance, the resistivity of copper is about 0.0172 μΩ for a 1 metre cube, and is therefore expressed as 0.0172 μΩm. The symbol for resistivity is the Greek letter ρ (pronounced 'rho'), which can now take the place of r in our previous formula, so that

$$R = \frac{\rho l}{a}$$

where R is the conductor resistance, ρ is the cable resistivity, l is the cable length and a is the cable cross-sectional area.

Put simply:

- After increasing the cable length, resistance will increase.
- After decreasing the cable cross-sectional area, resistance will increase.
- The reverse is true in each case.

Table 4.1 gives the resistivities of some common metallic conductors. These are average values, since a slight variation is found to occur depending on conductor condition.

Differing values of resistivity for different conductor materials make it clear that the resistance of a wire or cable depends on the material of which it is made, as well as on its length and cross-sectional area. Table 4.1 thus shows clearly why copper is so widely used as a conductor when the material temperature increases, being second only to silver in its resistivity value.

Aluminium is about half as resistive again as copper, but its lightness and cheapness have led to its increasing use for cables and busbars.

Table 4.1   Resistivity of common conductors

Material	Resistivity		
	μΩm	μΩcm	μΩmm
Copper (annealed)	0.0172	1.72	17.2
Copper (hard-drawn)	0.0178	1.78	17.8
Aluminium	0.0285	2.85	28.5
Tin	0.114	11.4	114
Lead	0.219	21.9	219
Mercury	0.958	95.8	958
Iron	0.100	10.0	100
Silver	0.0163	1.63	16.3
Brass	0.06–0.0	6–9	60–90

4.4 Resistance and resistivity calculations

All resistance calculations are based on the formula below (also given in Section 4.3).

$$R = \frac{\rho l}{a}$$

If resistivity, length and cross-sectional area are known, the resistance may be calculated. In fact, if any three of the four values are known, the fourth may be found.

Example 4.5

Calculate the resistance of 1000 m of 16 mm^2 single copper conductor.

ρ of copper $= 17.2 \times 10^{-8}$
$l = 1000$ m
$a = 16$ mm^2 convert to m^2 (SI Unit for area) $= 16 \times 10^{-6}$

$$R = \frac{\rho l}{a} = \frac{(17.2 \times 10^{-8}) \times 1000}{(16 \times 10^{-6})} = 1.075 \; \Omega$$

Example 4.6

Calculate the resistance of 10 m of conductor, if ρ for the material is 1.77×10^{-8} Ωm and the cross-sectional area of the conductor is 4.0 mm^2.

$$R = \frac{\rho l}{a} = \frac{(1.77 \times 10^{-8}) \times 10}{(4 \times 10^{-6})} = 0.044 \; \Omega\text{m}$$

Example 4.8

The heating spiral of an electric fire is removed from its silicon tube, uncoiled and measured. It has a length of 4 m, a diameter of 0.2 mm and its resistance is measured as 20 Ω. What is the resistivity, in Ωm, of the spiral material?

$$R = \frac{\rho l}{a} \quad \text{once transposed} \quad \rho = \frac{Ra}{l}$$

$$a = \frac{\pi d^2}{4} = \frac{3.142 \times 0.2^2}{4} = 0.03142 \text{ mm}^2$$

$$\frac{20 \times 10^6 \times 0.03142 \; \mu\Omega\text{mm}}{4000} = 157.1 \; \mu\Omega\text{mm}$$

$$p = \frac{20 \times (0.2 \times 10^{-6})}{4} = 1 \times 10^{-5} \; \Omega\text{m} \quad \text{or} \quad 0.000001 \; \Omega\text{m}$$

The examples above show the steps to be followed in this type of problem, which has many practical applications. The golden rule is to be careful to express resistivity, length and cross-sectional area in terms of the same units of area and length.

4.5   Effect of temperature on resistance

The following test will show that change of temperature can affect the resistance of a conductor.

When connected to a 2 V supply, a 60 W, 230 V lamp takes a current of 26.1 mA. Using Ohm's law, its resistance can be calculated:

$$R = \frac{U}{I} = \frac{2}{0.0261} = 76.6 \ \Omega$$

If the lamp is now connected to a 230 V supply, its filament becomes white hot and glows brightly and will be found to take a current of 261 mA. Its new resistance will be:

$$R = \frac{U}{I} = \frac{230}{0.261} = 881 \ \Omega$$

We can see that the increase in temperature has increased the filament resistance by 11.5. This extreme example is striking proof that temperature affects resistance.

It is often important to be able to calculate the resistance of a conductor at any given temperature. To be able to do this simply, we use the temperature **coefficient** of resistance for the material concerned, which is given the symbol α (Greek 'alpha').

KEY TERM: Coefficient – A fixed value multiplied by a variable.

The temperature coefficient of resistance of a material at 0 °C is the change in resistance of a 1 Ω sample of a given material, when its temperature is increased from 0 °C to 1 °C.

For instance, an **annealed** copper conductor that has a resistance of 1 Ω at 0 °C will have a resistance of 1.0043 Ω at 1 °C. Therefore we can say that the temperature coefficient of resistance of annealed copper at 0 °C is:

$$1.0043 - 1 = 0.0043 \text{ ohm per ohm per degree Celsius, or } 0.0043/^{\circ}\text{C}.$$

KEY TERM: Annealed copper – Copper that has been made softer and easier to work with through heat treatment.

Table 4.2   Temperature coefficients of resistance of some conductors

Material	(/°C at 0 °C)	(/°C at 20 °C)
Annealed copper	+0.0043	+0.00396
Hard-drawn copper	+0.0043	+0.00396
Aluminium	+0.0040	+0.00370
Brass	+0.0010	+0.00098
Iron	+0.0066	+0.00583
Nickel–chromium	+0.00017	+0.000169
Carbon (graphite)	−0.0005	−0.00047
Silver	+0.0041	+0.00379

In practice, it is often difficult to measure resistances at 0 °C, so temperature coefficients of resistance are often expressed for other temperatures, such as 20 °C. Table 4.2 gives some temperature coefficients of resistance for some common conductors.

Carbon has a negative temperature coefficient of resistance, which means that its resistance decreases as temperature increases.

Provided that the temperature coefficient of resistance of a conductor material is known, a simple formula can be deduced for calculating the conductor resistance at any temperature from the resistance at 0 °C.

- Let R_0 be the resistance of a conductor at 0 °C, and let α be its temperature coefficient of resistance in '/°C' at 0 °C.
- From the definition of temperature coefficient of resistance, the change in resistance will be R_0 ohms at 1 °C, $2R_0$ ohms at 2 °C and tR_0 ohms at $t°C$, if Rt is the total conductor resistance at $t°C$.

$$R_t = R_0 + R_0\alpha t$$

or

$$R_t = R_0(1 + \alpha t)$$

If the temperature coefficient of resistance at, say, 20 °C were being used, the formula could be written as $R_t = R_{20}(1 + \alpha t)$, where:

- R_{20} is the conductor resistance at 20 °C
- t is the change in temperature from 20 °C
- α is the temperature coefficient of resistance at 20 °C

Example 4.9
The resistance of 1000 m of 2.5 mm^2 annealed copper conductor is 5.375 Ω at 20 °C.
  Find its resistance at 50 °C.
  From Table 4.2, α is 0.00396/°C at 20 °C; therefore

$$R_{50} = R_{20}(1 + \alpha t)$$
$$= 5.375[1 + (0.00396 \times (50-20))]\Omega$$
$$= 5.375 \times 1.1188 = 6\ \Omega$$

One method of temperature measurement is to subject a resistor to the unknown temperature and accurately measure its resistance, hence calculating the temperature from a knowledge of the temperature coefficient of resistance of the resistor material.

Example 4.10
A resistance thermometer with a temperature coefficient of 0.001/°C at 0 °C and a resistance of 3 Ω at 0 °C is placed in the exhaust gases of an oil-fired furnace. The thermometer resistance rises, reaching a final steady value of 5.25 Ω. What is the exhaust-gas temperature?

$$R_t = R_0(1 + \alpha t)$$
$$5.25 = 3(1 + 0.001t)$$
$$5.25 = 3 + 0.003t$$
$$5.25 - 3 = 0.003t$$

Therefore,

$$t = \frac{2.25}{0.003} = 750\ °C$$

There are many practical cases where the temperature coefficient of resistance of a material is known at a given temperature, but where it is impracticable to measure the actual resistance of some particular conductor at this temperature.
  If R_1 is the conductor resistance at temperature t_1, and R_2 the resistance at temperature t_2,

$$R_1 = R_0(1 + \alpha t_1) \text{ and } R_2 = R_0(1 + \alpha t_2)$$

Dividing these equations, we have

$$\frac{R_1 = R_0(1 + \alpha t_1)}{R_2 = R_0(1 + \alpha t_2)} = \frac{1 + \alpha t_1}{1 + \alpha t_2}$$

Therefore,

$$R_2 = \frac{R_1(1 + \alpha t_2)}{1 + \alpha t_1}$$

R_0, the resistance at 0 °C, is not needed.

4.6   Effect of temperature changes

There are some applications, such as the resistance thermometer (see Example 4.10) where a resistive conductor is deliberately subjected to temperature changes. They are often made from materials that undergo a considerable change in resistance for a small temperature change. These resistors are called thermistors. Thermistors are used in temperature-measuring equipment, and in conjunction with electronic circuits in certain types of thermostat (Figure 4.8).

The change in resistance of a filament lamp described at the beginning of Section 4.5 illustrates how the change in resistance may cause problems. Taking the 60 W, 230 V lamp mentioned, the cold resistance is 76.6 Ω. At the instant of switching on the 230 V supply, the current to the lamp will be given by

$$I = \frac{U}{R} = \frac{230}{76.6} = 3 \text{ A}$$

At this instant, the lamp will be taking about 11.5 times its normal current. The lamp quickly heats up, increasing in resistance as it does so and reducing the current to its normal value of 261 mA.

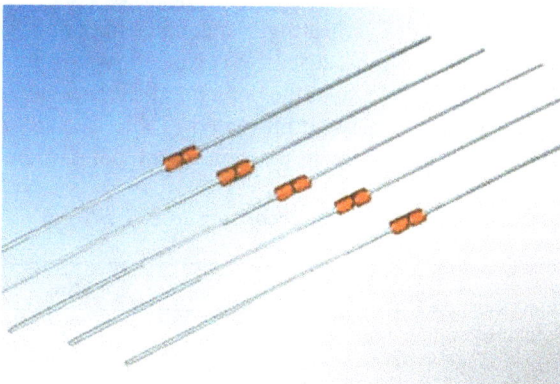

Figure 4.8   Thermistor

The short-lived heavy current, called a transient, can cause difficulties. The magnetic forces set up within the lamp itself may cause it to fail at the instant of switching.

This effect is not often noticeable because of the small currents taken by filament lamps. It would be serious, however, if the element of an electric fire were wound with tungsten, which is the material used for lamp filaments. The normal running current of a 1 kW fire element connected to a 230 V supply is about 4.35 A but would be about 50 A at the instant that a tungsten element is switched on. A fire element takes some time to reach its final temperature, so the heavy current would pass through the circuit for a period long enough to operate the circuit's protective device.

To prevent this, many conventional elements of heavily loaded heating equipment, such as fires, cookers and immersion heaters, are wound with a special alloy of nickel and chromium that has a near-zero temperature coefficient of resistance. This means that there will be almost no change in the resistance of the element as its temperature increases, and hence virtually no change in the circuit current.

There are many circumstances in which changes in the temperature of electrical equipment will affect its operation. For example, the windings of an electric motor will increase in temperature when it is used. This results in increased winding resistance and a reduction in the current. This may affect the operation of the motor. For example, reduction of current in the field winding of a direct-current motor will result in an increase in speed.

4.7   Voltage drop in cables

The resistance of cables will increase with increasing temperature, giving an increased cable-voltage drop. Usually, however, the cable resistance will form a very small proportion of total circuit resistance, and the effect will be unimportant. The increase in cable resistance when carrying fault current, however, may have a significant effect on fault loop impedance and on the safety of people using an electrical installation (see Chapter 13).

When choosing cables for an installation, it is necessary to ensure that they will carry the load current without overheating. An equally important factor that must be considered is the voltage drop caused by current flowing through their resistance. If this voltage drop is excessive, the potential difference across the load will be low and efficient operation may be prevented (Table 4.3). *BS 7671* states the allowable voltage drop from the supply position to any point in the installation.

Table 4.3   Maximum volt drop values according to IET BS7671

	Lighting	**Other uses**
Low-voltage installations supplied directly from a public low-voltage system	3%	5%
Low-voltage installation supplied from a private low-voltage supply	6%	8%

Since any voltage can be expressed by multiplying a current by a resistance ($U = IR$), voltage drop in a cable depends on the current it carries and on its resistance. Resistance depends on:

- conductor material
- cable length
- cross-sectional area
- temperature (see Section 4.3)

The resistance of a given length of cable and its voltage drop for a given current can be reduced by replacing it with a cable that has a larger cross-sectional area.

Example 4.12

An existing twin 1.0 mm² steel wire armoured cable with copper conductors is utilised by someone unskilled in electrical work to feed a 25 A heater. The cable is 18 m long.

Calculate

(a) The cable's voltage drop.
(b) The potential difference (PD) across the heater if the supply voltage is 230 V.
(c) The minimum cross-sectional area (CSA) of a replacement cable if the voltage drop is not to exceed 6 V?

(a) Cable voltage drop

- The resistivity (ρ) of the copper conductors (see Table 4.1) is 1.77×10^{-8} Ωm.
- The cross-sectional area of each conductor is 1.0 mm².
- The total conductor length will be 36 m:
    - 100 m line to the load
    - 100 m neutral back to supply).

$$\text{Cable resistance } R = \frac{\rho l}{a} \quad R = \frac{\left(1.77 \times 10^{-8}\right) \times 36}{\left(1 \times 10^{-6}\right)}$$

$$R = 0.64 \Omega$$

Cable volt drop $U = I \times R$ $\quad U = 25 \times 0.64 = 15.93$ V

(b) PD across heater

$$230 - 15.93 = 214.07 \text{ V}$$

$$\text{Volt drop} = \frac{214.07}{230} \times 100 = 93.07\% \quad 100 - 93.07 = 6.03\% \text{ volt drop}$$

Because maximum volt drop allowed for general power circuits = 5%, 6.03% is too high. The operation of the heater would be adversely affected by this drop in voltage because it is far outside the volt drop parameters set by BS7671.

(c) Minimum cross-sectional area (CSA) of a replacement cable

A voltage drop of 15.93 V is in excess of the allowable limit of 5% (11.5 V). If the voltage drop must not exceed 11.5 V, the maximum, the cable resistance should be:

$$R = \frac{U}{I} \quad \frac{11.5}{25} = 0.46 \ \Omega$$

$$R = \frac{\rho l}{a} \quad \text{so} \quad a = \frac{\rho l}{R}$$

$$a = \frac{(1.77 \times 10^{-8}) \times 36}{0.46} = 1.35 \ \text{mm}^2$$

This is the minimum CSA to satisfy the voltage-drop requirements of BS7671, but it is not a standard size. The next standard size above this is 1.5 mm² steel wire armoured cable that, according to Appendix 4 of BS7671, will carry a minimum of 27 A. This cable will carry 25 A; however, this is close and under perfect conditions. It is more likely that a 2.5 mm² cable would be used to provide a larger margin of error.

This exercise illustrates the importance of considering voltage drop as well as current-carrying capacity when choosing cables. Tables of cable current ratings in BS7671 include figures for the voltage drop per ampere per metre length of run for the cable concerned under the conditions indicated.

The BS7671 tables are more accurate than the method in the above examples because they take into account the installation method and cable reactance as well as resistance.

Reactance and its combination with resistance will be considered in Chapter 10. In some circumstances, it may be necessary to calculate the resistance of a supply cable to find its voltage drop. This process is illustrated in the following examples.

Example 4.13
A motor takes 45 A from a 230 V supply. The motor is fed from the supply by a twin copper cable 20 m long, each core of the cable has a cross-sectioned area of 10 mm². Calculate the voltage at the motor terminals.

Step 1 – Find the cable resistance
The total length of conductor (line and neutral) is 2 × 20 m = 40 m
Table 4.1 gives the resistivity of aluminium as 26.5 × 10⁻⁸ Ωm.

$$R = \frac{\rho l}{a} = \frac{(1.77 \times 10^{-8}) \times 400}{(10 \times 10^{-6})} = 0.071 \ \Omega$$

Step 2 – Calculate volt drop

Voltage drop $= I \times R = 45 \times 0.071 = 3.19$ V

Motor voltage = supply voltage − voltage drop = 230 − 3.19 V = 226.81 V
10 mm^2 is acceptable for this load because its volt drop is within the 11.5 V maximum.

Example 4.14
An industrial heater is fed from the supply by a 2.5 mm^2 copper single cables. The heater current is 20 A. The supply voltage is 230 V. If the potential difference at the heater terminals is 225.5 V, calculate the length of the supply cable to the nearest metre.

Step 1 – Find volt drop

Volt drop $= 230 - 225.5$ Volts $= 4.5$ V

$$\text{Cable resistance} = \frac{\text{Voltage drop}}{\text{Current}} \quad \frac{4.5}{20} = 0.225$$

$$R = \frac{\rho l}{a} \quad a = \frac{Ra}{\rho} \quad l = \frac{0.225 \times (2.5 \times 10^{-6})}{(1.77 \times 10^{-8})} = 31.87 \text{ m}$$

This is the total conductor length for the two cables (line and neutral). The actual cable length is half the conductor length, which to the nearest metre is 16 m.

4.8   Summary of formulas for Chapter 2

$$R = \frac{\rho l}{a} \quad a = \frac{\rho l}{R} \quad \rho = \frac{Ra}{l} \quad l = \frac{Ra}{\rho}$$

where R = conductor resistance, Ω; ρ = conductor resistivity, Ωm; l = conductor length, m; a = CSA, m^2.

$$R_t = R_0(1 + \alpha t) \quad R_0 = \frac{R_t}{1 + \alpha t} \quad \alpha = \frac{R_t - R_0}{R_0 t} \quad t = \frac{R_t - R_0}{R_0 \alpha}$$

where R_t = resistance at t °C; R_0 = resistance at 0 °C; α = temperature coefficient of resistance, °C, at 0 °C.

$$R_2 = \frac{R_1(1 + \alpha t_2)}{1 + \alpha t_1}$$

where R_1 = resistance at t_1 °C; R_2 = resistance at t_2 °C.

Cable voltage drop $U = IR$

where U = voltage, V; R = total cable resistance, Ω; I = cable current, A.

$$U_L = U_S - U$$

where U_L = voltage across the load terminals; U_s = supply voltage; U = cable voltage drop.

4.9   Exercises

1. The resistance of 100 m of 4.5 mm² cable is 0.36 Ω. What is the resistance of 600 m of this cable?
2. A single-core cable, 24 m long, has a measured conductor resistance of 0.06 Ω. What is the resistance of 1000 m of this cable?
3. The resistance of 15 mm² single cable is 1.12 Ω/km. Calculate the resistance of a twin 15 mm² cable, 80 m long, if the cores are connected in series.
4. A certain cable has a resistance of 0.5 Ω. What is the resistance of a cable made of the same material which has twice the cross-sectional area and is three times as long as the first?
5. Two cables have equal resistances, but one has a cross-sectional area 2.5 times greater than the other. How much longer is the thicker cable than the thinner cable?
6. How does the electrical resistance of a copper conductor vary with:
    (i)   an increase in its length
    (ii) an increase in its cross-sectional area
    (iii) an increase in its temperature?
        (a) Place the following materials in ascending order of resistivity: glass, copper, iron, carbon, brass.

7. A twin cable with aluminium conductors and a cross-sectional area of 25 mm² is 150 m long. What is the resistance of the cable? The resistivity of aluminium is 26.5×10^{-8} Ωm.
8. Calculate the resistance of 1000 m of copper conductor of 4 mm² cross-sectional area. Take the resistivity of copper as 1.77×10^{-8} Ωm.
9. An electric-fire element has a total length of 10 m, a cross-sectional area of 0.5 mm² and a resistance of 50 Ω. Calculate the resistivity of the material from which it is made.
10. Calculate the length of a single aluminium conductor of cross-sectional area 125 mm² that has a resistance of 0.1 Ω. The resistivity of aluminium is 26.5×10^{-8} Ωm.
11. The shunt-field winding of a DC motor has a resistance of 100 Ω at 20 °C. What will be its resistance at 45 °C, if it is made of copper ($\alpha = 0.00396$/°C at 20 °C)?

12. The resistance of 1 mm^2 copper conductor is 13 Ω/km at 20 °C. What will be its resistance/km at 60 °C ($\alpha = 0.00396/°C$ at 20 °C)?

13. A resistance thermometer has a temperature coefficient of 0.002/°C at 0 °C and a resistance of 40 Ω. When used to measure the temperature of an oven, the resistance increases to 56 Ω. What is the temperature of the oven?

14. A motor is used to drive a conveyor feeding a stove-enamelling plant, and its temperature when idle never falls below 30 °C. At this temperature, the winding resistance is measured as 15 Ω. When running, the operating temperature of the motor is 65 °C. What is the winding resistance at the higher temperature? The winding is of copper with $\alpha = 0.0043/°C$ at 0 °C.

15. The maximum permissible voltage drop in an installation is 4% of the supply voltage. What will be the minimum load voltage, to the nearest volt, for supply voltages of
    (a) 200 V
    (b) 230 V
    (c) 240 V
    (d) 250 V

16. Each core of a twin cable carrying a current of 10 A has a resistance of 0.15 Ω. What is the total voltage drop in the cable?

17. An industrial heater is connected to a 230 V supply by means of a twin cable, each core having a resistance of 0.08 . The heater takes a current of 25 A from the supply. What is the PD across its terminals?

18. What must be the maximum resistance of each core of a twin cable, feeding a heater taking 12 A from a 230 V supply, if the allowable voltage drop of 4% of supply voltage must not be exceeded?

19. A laboratory socket outlet is fed through a twin cable with each core having a resistance of 0.2 Ω. The supply is at 115 V, and the cable voltage drop must not exceed 1.5% of this value. What maximum current should be taken from the outlet?

20. A motor takes 10 A from a 200 V supply, and is fed through a twin 2.5 mm^2 copper cable 30 m long. Calculate the voltage at the motor terminals. ρ for copper is 1.77×10^{-8} Ωm

4.10 Multiple-choice exercises

4M1 The voltage drop in a cable carrying a direct current is due to its
    (a) impedance
    (b) resistance
    (c) size and type
    (d) reactance

4M2

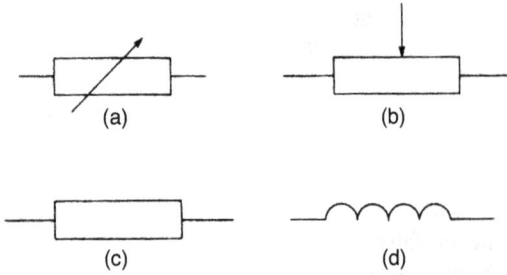

4M3 The resistance of a conductor is
   (a) directly proportional to its length
   (b) independent of its dimensions
   (c) directly proportional to its cross-sectional area
   (d) small

4M4 The type of resistor most likely to be found in a radio receiver is a
   (a) high-power woven-mat type
   (b) wire-wound encapsulated type
   (c) vitreous-enamelled wire-bound type
   (d) carbon low-power type

4M5 The resistor shown in the diagram is being used as a
   (a) potential divider
   (b) fixed-value resistor
   (c) heater
   (d) rheostat

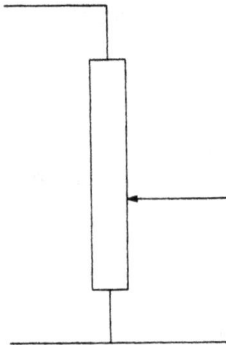

4M6 The resistance of a certain type of cable is 4 Ω/km. The resistance of 80 m of
   the same cable will be
   (a) 32 Ω
   (b) 50 Ω
   (c) 0.32 Ω
   (d) 0.5 Ω

4M7 Resistivity is usually quoted in
   (a) ohms
   (b) Ωm
   (c) microhms
   (d) rho

4M8 A conductor has a cross-sectional area of 20 mm^2, a length of 48 m and is made of a material with a resistivity of 0.02 Ωm. Its resistance is
   (a) 48 kΩ
   (b) 48 Ω
   (c) 0.48 Ω
   (d) 0.048 Ω

4M9 If a carbon resistor increases in temperature its resistance will
   (a) decrease
   (b) be unchanged
   (c) increase
   (d) fall to zero

4M10 The temperature coefficient of resistance of a conductor
   (a) indicates the maximum temperature that a resistor may reach
   (b) allows the calculation of resistance as temperature changes
   (c) shows that a resistor will get hot when it carries current
   (d) shows that its resistance will fall as it gets hotter

4M11 The current to a tungsten-filament lamp at the moment of switch-on
   (a) will blow the circuit fuse
   (b) is lower than the steady running current
   (c) usually causes the filament to burn out
   (d) is several times greater than after a second or so

4M12 The maximum voltage drop permitted in a circuit by BS 7671 is
   (a) 0
   (b) 0.4%
   (c) 5%
   (d) 12 V

Chapter 5

Mechanics

At first sight, it may seem strange that a book dealing with electrical theory should concern itself with mechanics. What connection is there? As well as understanding the operation of electrical equipment, an electrician or electrical engineer may have to manufacture as well as install it. This involves various mechanical operations, such as cutting cables, threading conduits, driving screws, lifting heavy apparatus and so on. An understanding of the principles involved might reduce both the physical effort required and the likelihood of an accident occurring while carrying out these tasks.

5.1 Mass, force, pressure and torque

Mass (symbol, m; measurement, kg)

Mass is the amount of material in an object, and is usually measured by comparison with another mass chosen as a standard. For instance, the standard of mass in SI units is the kilogram (kg), which is the mass of a block of platinum kept at Sèvres in France.

Force (symbol, F; measurement, newtons N)

Force can be measured in terms of the effects it produces. A force can lift, bend or break an object. It can move an object that was previously stationary, or can stop or change the rate and direction of movement. In electrical craft practice, force is used to move, bend, cut or join together materials, as well as to drive in screws and perform a host of other tasks.

Velocity (symbol, v; measurement, metres/second m/s)

Velocity is the rate that an object changes position. Velocity has the same unit of measure as speed, which is metres/second. Unlike speed, however, when calculating velocity the *direction* of movement has to be considered. This is called the vector quantity.

Weight

The force exerted by an object due to gravity. It can be expressed as $F = mg$

- m – mass kg
- g – Earth's gravity, which exerts a force of 9.81 N/kg

Also, the velocity achieved by a falling object due to Earth's gravity is 9.81 m/s^2. In other words, an object will increase its rate of fall velocity by 9.81 m/s every second.

The Earth exerts a natural force of attraction on all other masses. This force is generally called the 'gravitational pull' of the Earth. The magnitude of this force acting on an object is generally considered to be the object's weight. Weight, therefore, is not a **constant**. In space, for example, where there is little or no gravitational force acting on the object, its weight is virtually zero. On the moon, gravitational force is 1.64 N/kg, because the gravitational force on the moon is approximately 0.17× that on Earth.

Scientists and engineers require a constant to calculate force, power and energy so an object's mass is used rather than its weight, because mass is a constant. An object of 500 kg mass on Earth will also have a mass of 500 kg in space or on the moon.

> **KEY TERM: Constant – A value that is not changed by external conditions. Pi, for example, is a constant and is used as a value to calculate properties of a circle. Pi is always 3.142 (to the first three decimal places), no matter the size of the circle.**

Torque (symbol, τ; measurement, newton-metre Nm)

In some cases, a force may cause rotary or turning movement. Probably the simplest examples of this principle in electrical work are tightening a screw using a screwdriver or threading conduit with a stock and die. The turning effect is called the torque, or sometimes, the turning moment of a force.

It is measured as the force applied $(F) \times$ the perpendicular distance between the direction of the force and the point about which the force is rotated – in other words, the radius (r) (see Figure 5.1):

$$t = Fr$$

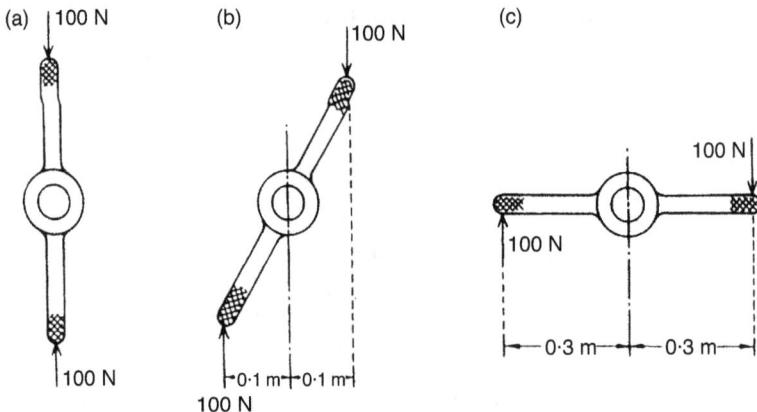

Figure 5.1   Torque applied to stocks and dies: (a) torque = 100 N ×0×2 = 0; (b) torque = 100 N × 0.1 m × 2 = 20 Nm and (c) torque = 100 N × 0.3 m × 2 = 60 Nm

Figure 5.1 makes it clear that to obtain maximum torque for a given applied force, the force should always be in the direction of the resulting movement.

Torque (or turning moment) is measured in newton metres (Nm).

- In Figure 5.1(a), the distance between the direction of the force and the turning point is zero. There is no torque despite the application of a force of 100 N This force is trying to compress the stock and not to rotate it.
- The torque produced in Figure 5.1(b) will produce movement, but due to the steep angle some of the force will still be wasted trying to distort the stock.
- In Figure 5.1(c), the distance between the direction of the force and the turning point is maximum. In other words, all the force applied is being used to turn the stock, and none is wasted in trying to distort it.

The actual torque required to thread a conduit depends on its size, but can be reduced considerably by the use of:

- a sharp die
- correct lubrication.

For a given torque, the force required can be reduced if exerted at a greater radius from the turning point (Example 5.1).

Example 5.1

A conduit requires a torque of 20 Nm for the threading operation. What minimum force must be exerted by each hand pressing at the ends of a stock of overall length:

(a) 0.5 m
(b) 1 m?

Each hand provides half the total torque, so the torque per hand will be 10 Nm

Torque = Force direction × Distance, so by transposition, Force

$$= \frac{\text{Torque}}{\text{Distance}}$$

In both cases, the direct distance from the point of application of the force to the turning point will be half the overall length of the stock.

(a) Minimum force required = $\frac{10 \text{ Nm}}{0.25 \text{ m}}$ = 40 N

(b) Minimum force required = $\frac{10 \text{ Nm}}{0.5 \text{ m}}$ = 20 N

An electrician should not be tempted the to increase the effective length of their tool handles using a suitable length of pipe or other means. The excessive torque that can then be applied could damage tools. A tool will usually be made with handles of such length as to prevent the application of too much torque by a person of normal strength.

5.2   Units of force and weight

A mass of 1 kg will experience a force due to the Earth's gravity of 1 N. However, since the kilogram is a unit of mass, it should not be used to measure force. A mass of one kilogram experiences a force due to gravity of 9.81 N/kg.

We could say, for example, that a piece of machinery has a mass of 1 kg, in which case it will require a force of 9.81 N to lift it against gravity. If the machinery formed part of a space rocket, it could be sent outside the gravitational pull of the Earth. Its mass would then be unchanged, but it would experience none of Earth's gravitational force.

Example 5.2
A bundle of conduit has a mass of 800 kg. What force will be needed to lift it?
One mass of 1 kg requires a force of 9.81 N to lift it against gravity.

Force required = $800 \times 9.81 = 7848$ N

Pressure (symbol, p; measurement, pascals (Pa) or newton/ metres^2 Nm^2)

Pressure on a surface is measured in terms of the force per unit area on that surface, assuming that the surface is at right angles to the direction of the force. For example, a force of 500 N acting on one leg of a tripod pipe vice with an effective area of 20 mm^2 in contact with a floor will exert a pressure given by:

$$\text{Pressure} = \frac{\text{Force } (N)}{\text{Effective area (mm}^2)} = \frac{500}{20} = 25 \text{ Pa}$$

A pressure this high as this will obviously damage many floor finishes. However, if the leg of the vice rests on a piece of wood with an area of 100 mm^2, the same force will be spread over an area of 100 mm$^2 \times$ 100 mm$^2 =$ 10,000 mm^2.

$$\frac{500}{10,000} = 0.05 \text{ Pa}$$

Note that in the example, the term 'effective area' was used. Normally, the SI unit for area is m^2, not mm^2. In this case, however, we are looking for the pressure exerted on the actual area of contact between the tripod's foot and the floor, 20 mm^2 and then 10,000 mm^2.

Density (measured in kg/m^3)

If we take two blocks of the same size, one made of wood and the other of iron, the iron block will be heavier than the wooden block. This is because iron is denser than wood. The density of a material is the mass of that material contained in one cubic unit, and is given in kilograms per cubic metre (kg/m^3).

Example 5.3

What is the mass of a block of copper 0.1 m by 0.2 m by 0.15 m? The density of copper is 8900 kg/m³.

Find volume of the block:

$$0.1 \times 0.2 \times 0.15 = 0.003 \text{ m}^3$$

Now find the mass

$$\text{Density} = \frac{\text{Mass}}{\text{Volume}} \quad \text{Mass} = \text{Density} \times \text{Volume}$$

$$= 8900 \times 0.003 \text{ kg}$$

$$= 26.7 \text{ kg}$$

5.3 Work, energy and power

Work (symbol, w; measurement, joule J or Nm)

Work is done when a force is applied to a body and movement results. For instance, work is done when a weight is lifted or when force is applied to a screwdriver, causing it to rotate. You will notice that both work and energy are measured in joules. The difference between the two is that:

- Work is a form of applied energy, energy used to accomplish a physical task.
- Energy may be present, but it may not be doing any work at that moment.

Work is measured as the product of the force applied to move an object a certain distance.

When the movement is in the same direction as the force, the work done is equal to the force exerted × distance moved.

$$\text{Work} = \text{Force} \times \text{Distance} \quad w = Fd$$

Example 5.4

A bundle of conduit has a mass of 200 kg.

(a) What is its weight?
(b) What work must be done in lifting the conduit from the floor on to a rack 2 m high?

Remember, weight is the force exerted on the mass by gravity. A mass of 1 kg exerts a force due to gravity of 9.81 N.

(a) Weight

$$\text{Weight} = \text{Mass} \times 9.81 \quad \text{Weight of conduit} = 200 \times 9.81 = 1962 \text{ N}$$

(b) Work

$$w = Fd \quad w = 2 \times 1962 = 3924 \text{ J}$$

Example 5.5

A force of 100 N will just move a van on a level road. What work will be expended in pushing the van 15 m?

$$w = Fd = 15 \times 100 = 1500 \text{ J}$$

Since the van is not being lifted, its weight is only important in its effect on the force needed to move it against friction.

Energy (symbol, w; measurement, joules J)

Energy is the capacity to do work. It may take many forms, such as

- nuclear
- chemical
- thermal
- mechanical
- electrical

Energy can be classed as:

- Potential energy – energy that can be released but is currently stored within an object, machine or system. For example, a bolder on the edge of a cliff is not expending any energy, but if it falls, energy will be released.
- Kinetic energy – energy that is being expended, for example, the energies of movement and heat produced by the falling boulder.

If we ignore theoretical atomic physics, which never affects electrical-craft work, it is true to say that energy can neither be created or destroyed. It can, however, be converted from one form to another.

For example,

1. Gas containing chemical energy is burned in the boilers of a power station producing heat.
2. This heat evaporates water to become steam under pressure.
3. Steam is fed to a turbine where mechanical energy is produced.
4. The turbine drives an alternator which produces electrical energy.

Efficiency (symbol, η; measured as a percentage)

Although energy cannot be created or destroyed, it does not follow that energy can be converted from one form to another without waste. For instance, the conversions in a power station generally result in little more than one-third of the available chemical energy becoming electrical energy. The difference is dissipated largely in the form of heat, and is lost from the process. Improvements in design have reduced these losses considerably in recent years, which means that power stations have become more efficient.

The efficiency of a mechanical system can be defined as the ratio of output to input and is usually expressed as a percentage so that

$$\text{Efficiency } (\eta) = \frac{\text{Output energy or power}}{\text{Input energy or power}} \times 100$$

In practice, it is more usual to express efficiency in terms of power than energy.

The difference between input and output is the wasted energy, or losses, so that

Output = Input − Losses

Input = Output + Losses

Which means that

$$\text{Efficiency} = \frac{\text{Output}}{\text{Output} + \text{Losses}} \times 100$$

$$\text{Efficiency} = \frac{\text{Input} - \text{Losses}}{\text{Input}} \times 100$$

Power (symbol, P; measured in watts W)

Power is defined as the rate of doing work. An electrician is more familiar with power in terms of electricity usage, kW/h, for instance, used to measure an installation's energy consumption, or the power rating of an electric heater, which enables them to calculate its current rating and, from that, the required cable size. Power, however, can also be applied to mechanical work.

An electrician is using a hammer and chisel to knock a hole in a brick wall. The same hole could be cut more quickly with a pneumatic drill. In either method, the completed hole will represent the same amount of work, but the rate of doing work is greater in the second method because the pneumatic drill is more powerful than the electrician.

$$\text{Average power} = \frac{\text{Amount of work done}}{\text{Time taken to do it}} \text{ or } P = \frac{w}{t}$$

The unit is the joule per second (J/s), also known as the watt (W). The watt is too small for many practical applications, so the kilowatt (kW) is often used.

1 kW = 1000 W

1 kW = 1000 J/s

Example 5.6

An electric motor drives a pump which lifts 1000 l (litres) of water each minute to a tank 20 m above normal water level. What power must the motor provide if the pump is 50% efficient? One litre of water weighs 9.81 N.

Force required to lift water

SI unit for time is second so convert 1 min to seconds = 60 s

$$F = \frac{1000 \times 9.81}{60} = 163.5 \text{ N/s}$$

Power required

Power dissipated by pump = Rate of lifting water × distance water to be lifted

$$P = 163.5 \times 20 = 3270 \text{ W or } 3.27 \text{ kW}$$

Power required due to 50% pump efficiency

$$\text{Output power of motor} = 3270 \times \frac{100}{50} = 6540 \text{ W or } 6.54 \text{ kW}$$

5.4   Lifting machines

A lifting machine is a device that requires a small effort to lift a large load. Such machines are often necessary in electrical work to move heavy equipment. A few of the simpler types are considered here.

The lever

A lever is the basic method of expending a certain amount of force to overcome a greater amount of force. The simplest form of lever has a load close to the pivot point, or fulcrum, while the effort is applied at a further point on the opposite side of the fulcrum. This principle applies to a number of different, and sometimes more complex, forms of lever. For example:

- Spade or shovel
- Scissor or shears
- Hammer

The simplest force equation that can be applied to a lever is:

Load force × Load distance (from fulcrum) = Effort distance (from fulcrum) × Effort force

However, we also need to consider the torque involved in the operation of a lever.

Figure 5.2 shows a simple lever. The total turning moment or torque applied to one end of the lever must be equal and opposite to the torque available at the other end.

If 50 N of effort is applied to the right-hand end of the lever as shown, the total applied torque will be 50 N × 2 m or 100 Nm.

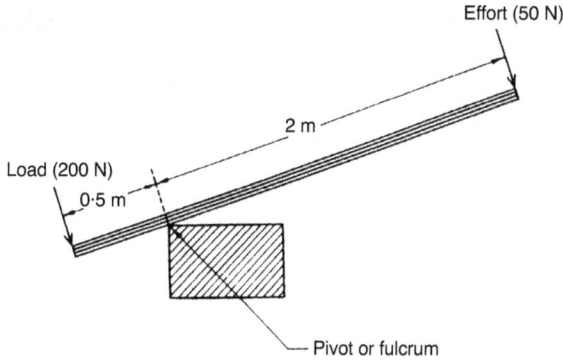

Figure 5.2   Simple lever

If there are no torque losses (and losses here will be very small), the same torque applied to the right-hand end of the lever will be available at the left-hand end. The radius of action about the pivot, fulcrum, is 0.5 m at this end.

Remember the torque calculation:

$$\tau = Fr \quad \tau = 50 \times 2 = 100 \text{ Nm}$$

To find the available force on the load side of the fulcrum

$$F = \frac{t}{r} \frac{100 \text{ Nm}}{0.5 \text{ m}} = 200 \text{ N}$$

This shows that this lever enables a load of 200 N to be balanced by a force of 50 N, because the lever is four times as long on one side of the fulcrum as it is on the other.

Levers are classified according to the relative positions of the load, the effort and the fulcrum, as shown in Figure 5.3.

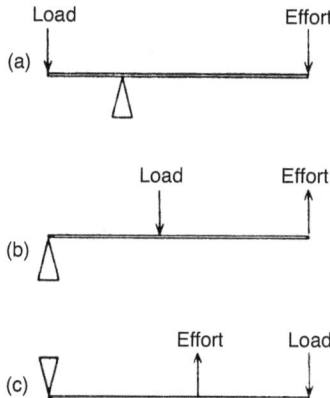

Figure 5.3   Classification of simple levers: (a) Class 1 lever; (b) Class 2 lever and (c) Class 3 lever. Image credit: MikeRun, via Wikimedia Commons, CC BY-SA 4.0.

Mechanical advantage and velocity ratio (symbol, mechanical advantage MA; velocity ratio, VR)

Mechanical advantage

Mechanical advantage is the ratio of load to effort.

$$MA = \frac{Load}{Effort}$$

The lever in Figure 5.2 has the advantage of allowing a large mass to be lifted by a small force.

$$MA = \frac{Force\ exerted\ by\ load}{Force\ exerted\ by\ effort} = \frac{200}{50} = 4\ or\ a\ 4:1$$

This is a ratio, so there is no unit.

There are factors to be considered when calculating mechanical advantage.

- For a simple lever, the mechanical advantage will be almost constant and for a **class 1 lever**, will depend on the lengths of the lever on either side of the fulcrum.
- For most machines, however, friction will vary with load, and mechanical advantage will not be constant.

   KEY TERM: Classes of lever

- **Class 1: the fulcrum is at a point between the effort and the load.**
   - **Class 2: The load is in the middle, between the fulcrum and the effort, e.g. a wheelbarrow.**
   - **Class 3: The effort sits between the fulcrum and the load, e.g. a human arm when bending from the elbow.**

Velocity ratio

Velocity ratio is defined as the ratio of the distances moved by the effort and the load:

$$VR = \frac{The\ distance\ moved\ by\ the\ effort\ in\ a\ given\ time}{The\ corresponding\ distance\ moved\ by\ the\ load}$$

For example, it will be clear from Figure 5.2 that, if the load is moved by 1 mm, the effort must move by 4 mm. This means that the velocity ratio will be 4 (no units apply). This is the same ratio as the mechanical advantage.

Note that this only applies to machines that are 100% efficient:

Work input = Work output

so

Effort × Distance moved by effort = Load × Distance moved by load

$$\frac{\text{Distance moved by effort}}{\text{Distance moved by load}} = \frac{\text{Load}}{\text{Effort}}$$

Velocity ratio = Mechanical advantage

For most machines, the inefficiency results in a smaller value for mechanical advantage than for velocity ratio, and it is interesting to note that

$$\text{Efficiency} = \frac{\text{Work out}}{\text{Work in}}$$

$$= \frac{\text{Load} \times \text{Distance moved by load}}{\text{Effort} \times \text{Distance moved by effort}}$$

$$= \frac{\text{Load}}{\text{Effort}} \div \frac{\text{Distance moved by effort}}{\text{Distance moved by load}}$$

$$= \frac{\text{Mechanical advantage}}{\text{Velocity ratio}}$$

Example 5.7
A class 1 lever is arranged so that a load of 1000 N is to be lifted at a distance of 200 mm from the fulcrum.

(a) What force must be applied 2 m from the fulcrum to balance the load?
(b) Calculate the mechanical advantage of the system, assuming no losses.
(c) Calculate the velocity ratio of the system, assuming no losses.

(a) Force
Let F be the required force.

$$\text{Convert 200 mm to m} = \frac{200}{1000} = 0.2 \text{ m}$$

Torque or turning moment required for load

$$\tau = Fr \quad \tau = 1000 \times 0.2 \quad \tau = 200 \text{ Nm}$$

If there is no waste, this will equal the turning moment to be applied. Therefore

$$F \times 2 = 200 \text{ Nm}$$

and

$$F = \frac{\tau}{r}\frac{200}{2} = 100 \text{ N}$$

(b) Mechanical advantage

$$MA = \frac{Load}{Effort} = \frac{1000}{100} = 10$$

(c) Velocity ratio

$$VR = \frac{Effort\ distance}{Load\ distance} = \frac{2}{0.2} = 10$$

Inclined planes and jacks

A simple method of lifting a heavy load is to push it up an inclined plane. For example, if a lorry is not fitted with a lift, it is easier to manhandle heavy machinery on to it using a number of planks to form an inclined plane from the ground to the lorry. If necessary, a box or trestle can be placed beneath the centre of the planks to give extra support.

The load is pushed up the sloping planks in a series of stages. This will involve more work than a direct lift, owing to the friction between the load and the planks, but the force required at any instant is less and rests may be taken between efforts. A rope, secured at one end of the load, round a solidly fixed structure such as a girder or the lorry chassis, is a useful safety measure to ensure that the load does not slip. Slack rope is taken up as the load is lifted.

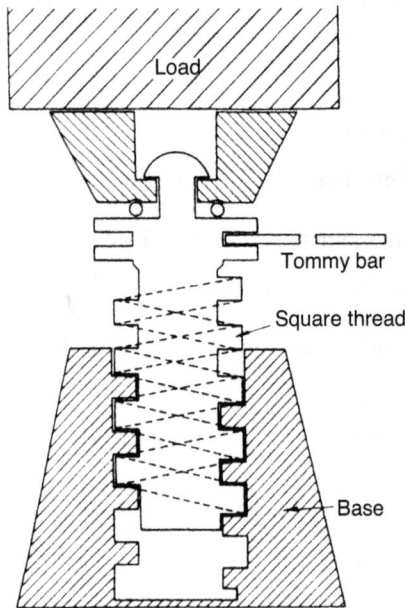

Figure 5.4   Simple Jack

In some machines, a steel inclined plane is, in effect, wound round a central column to form a screw thread. Rotation of the thread can lift a load by forcing it up the inclined plane. A machine of this sort is called a jack. One type of jack is shown in Figure 5.4.

Pulley systems

A block-and-tackle is a machine often used to lift heavy loads where a suitable overhead suspension is available (Figure 5.5). This consists of two sets of pulleys and a length of rope. A simple arrangement with two blocks each having two pulleys is shown in Figure 5.6. The pulleys in each block are usually the same size and are mounted side-by-side, but are drawn as shown for clarity.

Neglecting the rope to which the effort is applied, there are four ropes supporting the load.

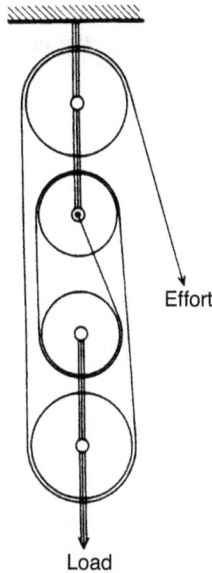

Figure 5.5   Principle of block and tackle

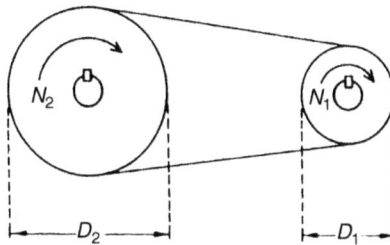

Figure 5.6   Belt and pulleys. Two pulleys coupled by a belt.

If the load is to be lifted by 1 m, each of these four ropes must shorten by 1 m, so that the effort rope is pulled through 4 m.

- The velocity ratio is, therefore, equal to four in this case, and in fact the velocity ratio of a block and tackle is equal to the number of pulleys used.
- The mechanical advantage (MA) would be equal to the velocity ratio (VR) if there were no losses, but friction at the pulley bearings reduces its value.

Example 5.8

A block and tackle of two sets, each of five pulleys, is 60% efficient. What is the maximum load that a man weighing 600 N could support using it?

$$\text{VR} = 2 \times 5 = 10$$

$$\text{MA} = \text{VR} \times \text{Efficiency} = 10 \times \frac{60}{100} = 6$$

Maximum effort will be applied when the man pulls the rope. This is an effort of 600 N.

$$\text{MA} = \frac{\text{Load}}{\text{Effort}}$$

$$\text{Load} = \text{MA} \times \text{Effort} = 6 \times 600 = 3600 \text{ N}$$

5.5   Power transmission

Belts and pulleys

One method of attaching a motor to a machine is to fit both with a pulley wheel and connect via a belt. Most belts are of the 'vee' type, which grip the pulley and reduce slip.

If the two pulleys are the same size, speeds and torques will be identical – if belt slip is neglected. If the pulleys are of differing sizes, it will result in a speed change between the large and smaller pulley. The belt, however will travel at a uniform speed.

Consider the arrangement shown in Figure 5.6.

- A motor of speed N1 revolutions per second is fitted with a pulley of diameter D_1 metres.
- The motor pulley is coupled by a belt to a machine pulley of diameter D_2 metres and having a speed of N_2 revolutions per second.
- The belt makes contact with both pulleys; so, if we neglect belt slip, the pulleys will have an identical edge, or peripheral, speed.
- The motor pulley has a circumference of πD_1 metres, and since it makes N_1 revolutions each second, its peripheral speed is $\pi D_1 N_1$ metres per second.

- Similarly, the peripheral speed of the machine pulley is $\pi D_2 N_2$ metres per second.

Since the peripheral speeds of both pulleys are equal if belt slip is neglected,

$$\pi D_1 N_1 = \pi D_2 N_2$$

so that

$$\frac{N_1}{N_2} = \frac{D_2}{D_1}$$

Thus, pulley speeds are in inverse ratio to their diameters. As well as changing the speed, pulleys of different sizes also change the torque. It would be wrong to say that no losses occurred in the belt-drive system, but if we neglect these losses, the work done by the motor is given to the machine it drives, so that the two are equal.

- Let the force exerted at the edge of the motor pulley be F_1 newtons.
- Let the machine pulley be F_2 newtons.
- In one revolution, the force on the motor pulley moves through πD_1 metres, so that the work done in 1 s is $\pi D_1 N_1 F_1$ joules.
- Similarly, the work done on the machine pulley in 1 s is $\pi D_2 N_2 F_2$ joules.

Assuming no losses, these values are the same so that

$$\pi D_2 N_2 F_2 = \pi D_1 N_1 F_1$$

or

$$D_2 N_2 F_2 = D_1 N_1 F_1$$

The torque on the motor pulley is $F_1(D_1/2)$, and that on the machine pulley is $F_2(D_2/2)$.

If motor torque is T_1 and machine torque is T_2,

$$T_2 N_2 = T_1 N_1$$

or

$$\frac{T_1}{T_2} = \frac{N_2}{N_1} = \frac{D_1}{D_2}$$

The pulley torques are in:

- inverse ratio to their speeds
- direct ratio to their diameters.

If a belt-and-pulley system is used to

- reduce speed, it must *increase* torque
- increase speed, it will *reduce* available torque.

Example 5.9
A motor providing a torque of 300 Nm at a speed of 24 r/s is fitted with a pulley of diameter 100 mm. This pulley is coupled by a belt to a machine pulley of diameter 400 mm. Assuming no losses or belt slip, calculate the:

(a) machine pulley speed
(b) torque it can provide.

(a) Pulley speed

$$\text{Convert mm to m} \ \frac{100}{1000} = 0.1 \text{ m and} \frac{400}{1000} = 0.4 \text{ m}$$

$$\frac{N_1}{N_2} = \frac{D_2}{D_1}$$

therefore

$$\frac{N_2}{D_1} = N_1 D_1 = 24 \times \frac{0.1}{0.4} = 6 \text{ r/s}$$

(b) Torque

$$\frac{T_1}{T_2} = \frac{D_1}{D_2}$$

therefore

$$\frac{T_2}{D_1} = T_1 D_2 = \frac{300}{0.1} \times 0.4 = 1200 \text{ Nm}$$

Gearing

Instead of using a belt to connect two pulley wheels, each wheel may have teeth cut in its edge. If the two sets of teeth are meshed, they will turn together and have the same peripheral speed. Toothed wheels of this sort are called gears. They are more positive than a belt as they cannot slip, but may be noisy and will usually require periodic lubrication.

Since peripheral speeds are identical, the same reasoning applies to gears as to pulleys. It should, however, be noted that two pulleys coupled by a belt, rotate in the same direction, whereas two meshed gears rotate in opposite directions.

Gear sizes are usually indicated by the number of teeth rather than by diameter, the two being proportional. So, if G_1 and G_2 represent the numbers of teeth on two gearwheels, respectively:

$$\frac{T_1}{T_2} = \frac{N_2}{N_1} = \frac{G_1}{G_2}$$

Example 5.10

A motor with a speed of 12 r/s provides a torque of 50 Nm, and is fitted with a gearwheel having 30 teeth. This wheel is meshed with a second gearwheel, which has 12 teeth. Calculate the:

(a) speed of the second gearwheel
(b) torque it provides.

(a) Speed of the second gearwheel

$$N_2 = \frac{N_1 G_1}{G_2} = 12 \times \frac{30}{12} = 30 \, \text{r/s}$$

(b) Torque provided by the second gearwheel

$$T_2 = \frac{T_1 G_2}{G_1} = 50 \times \frac{12}{30} = 20 \, \text{Nm}$$

Chains

A chain coupled to two or more rotating **sprockets** can be considered as a pulley-and-belt system and follows the same rules. The chain is more expensive and is noisier than the belt, but it cannot slip. It is used where a **positive drive** is needed.

> **KEY TERM: Sprocket** – A wheel with teeth cut into its rim designed to carry a chain. An example would be the sprockets that carry a bicycle chain.

> **KEY TERM: Positive Drive** – A drive in which there can be no slip.

Rotating shafts

As well as the drives, the simple rotating shaft is a common mechanical power transmitter. Without belts, gears or chains, there is no change in speed or torque with a shaft drive.

5.6 Parallelogram and triangle of forces

Scalars and vectors

Mechanical-engineering quantities can be classified as either scalars or vectors.

Scalars

Those that can be measured in terms of magnitude only are called scalar quantities. Examples of scalar quantities are:

- Length
- Mass

- Power
- Speed

Vectors

Other quantities can only be completely described if reference is made to direction as well as magnitude. These are called vector quantities. For example, the position of an aircraft after a given time can only be forecast accurately if not only its starting position and speed, but also its direction, are known. Velocity is a vector quantity.

A force is also a vector quantity and is often represented by a line called a vector. The length of the vector, drawn to a suitable scale, represents the magnitude of the force. The direction of the line is the same as that of the force it represents. An arrowhead is added to show the direction (push or pull) of the force, which is assumed to act at a point at one end of the vector.

Equilibrium

A body is in a state of equilibrium when it is at rest or in a state of uniform motion. For instance, a luminaire suspended from a chain attached to a ceiling is in equilibrium because it does not move. The downward acting force due to the weight of the fitting is opposed by the equal upward supporting force of the chain. If the weight of the fitting is too great for its fixings to bear, the downward force exceeds the upward force, equilibrium is lost and the fitting will fall.

Resultant of forces

It often happens that a body is acted on simultaneously by a number of forces. For simplicity, we can assume that these forces are removed and replaced by one force, called the resultant. The resultant will have exactly the same effect on the body as the forces it replaces.

The resultant of two forces can be found by completing the parallelogram of forces as shown in Figure 5.7. Two forces, **P** and **Q**, represented by the vectors OA and OB, have the resultant **R**, represented by the vector OC.

If a resultant for more than two forces is required, the resultant of two vectors are taken one at a time until the system is resolved into one vector only. It is often necessary to find the force which will just balance all the other forces acting on a

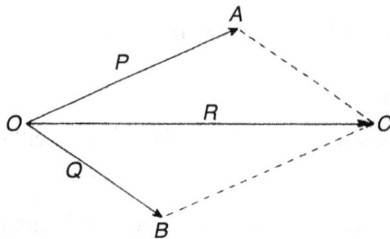

Figure 5.7   Parallelogram of forces. R is resultant of forces P and Q.

body. This is called the equilibrant, and is equal in magnitude and opposite in direction to the resultant of the forces.

Example 5.11

A pole supporting overhead cables is situated so that the cables meet it at a right angle. The cables in one direction exert a pull of 4000 N on the pole, and those in the other direction a pull of 3000 N. Find the

(a) horizontal pull on a stay wire
(b) direction in which it must be fixed so as to balance the horizontal forces exerted by the cables on the pole.

The arrangement of the system is shown in Figure 5.8.

- represents the pole, and vectors
- OA and OB represent the forces of 4000 N and 3000 N.

These vectors are drawn to scale and at right angles to each other. The parallelogram of forces is completed with the resultant vector OC representing a force of 5000 N in the direction shown.

The stay wire must be the equilibrant of the resultant OC, and is represented by the equal and opposite vector OD.

The horizontal pull on the stay wire will be 5000 N, and its direction, measured on the vector diagram with a protractor, will be approximately 143° from the 4000 N cables in an anticlockwise direction.

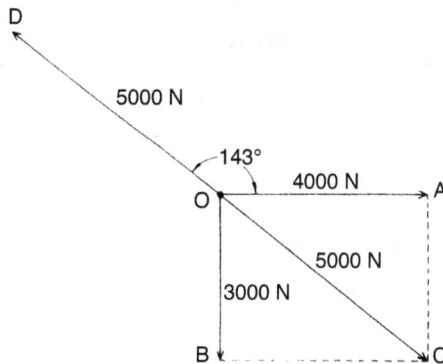

Figure 5.8 *Vector diagram for Example 5.11*

Triangle of forces

In Figure 5.9(a), the parallelogram of forces is used to find the resultant *C* and the equilibrant *D* of two forces *A* and *B*.

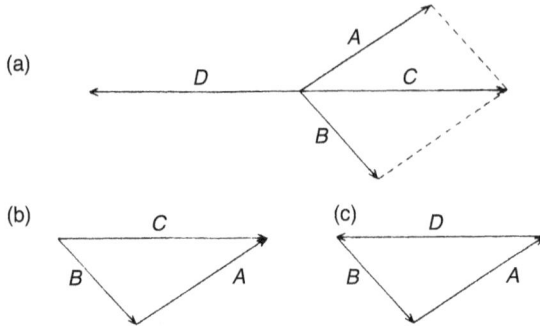

*Figure 5.9   Triangle of forces: (a) parallelogram of forces used to find resultant C
and equilibrant D of two forces A and B; (b) part of parallelogram of
forces used to find resultant of same two forces and (c) triangle of
forces used to find equilibrant of same two forces.*

Figure 5.9(b) shows a part of the parallelogram used to find the resultant **C**, and
illustrates a useful saving in space which could lead to a larger scale and improved
accuracy.

Figure 5.9(c) is the same as Figure 5.9(b), but with the equilibrant **D** substituted
for the resultant **C**. The three forces shown form the three sides of the triangle of forces.

Example 5.12

A fluorescent light fitting weighing 140 N is to be suspended from a single support
chain connected to two sling chains taken to the fitting ends as shown in Figure 5.10(a).
Find the tension in each sling chain, if each makes an angle of 30° with the fitting.

A simple figure, often called a space diagram (Figure 5.10(b)) will help us to
understand the problem. Since the fitting is in a state of equilibrium, so must be the
three forces in the space diagram, where:

- OA is the tension in the support chain.
- OB and OC are the tensions in the two sling chains.

Clearly, the tension in the support chain must be equal to the weight of the
fitting, and will be 140 N.

The triangle of forces is then drawn.

1. OA, to a scale length equivalent to 140 N, is drawn as shown in Figure 5.10(c).
2. Lines from points O and A, each making an angle of 60° with OA and meeting
   at point D to complete the triangle.

These lines represent the tensions in the sling chains, and when measured are
found to have a length which is equivalent to 140 N.

Although the sling chains support the weight between them, each has the same
tension as the support chain. Failure to appreciate the fact that the tension in the

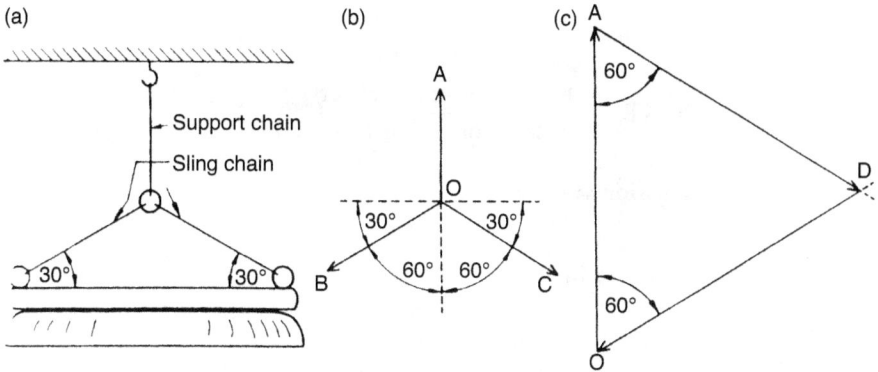

*Figure 5.10   Space diagram and triangle of forces of Example 5.12:
(a) fluorescent light fitting supported from sling chains; (b) space
diagram for arrangement of (a) and (c) triangle of forces for
arrangement of (a)*

sling chains increases as the angle they make with the fitting decreases has been
responsible for a number of accidents. Here, the sling angles are equal. If they were
unequal, the chain tensions would be different.

5.7   Summary of formulas for Chapter 3

A mass of 1 kg experiences a force due to gravity of 9.81 N.

$$\text{Pressure, N/m}^2 = \frac{\text{Force}}{\text{Effective area}} \qquad p = \frac{F}{a}$$

$$\text{Density, kg/m}^3 = \frac{\text{Mass}}{\text{Volume}} \qquad \rho = \frac{m}{L^3}$$

$$\text{Torque, Nm} = \text{Turning force} \times \text{Radius of action} \qquad \tau = Fr$$

$$\text{Work or energy, } J = \text{Force required} \times \text{Distance moved} \qquad w = Fd$$

$$\text{Power} = \frac{\text{Work or energy}}{\text{Time}} \qquad P = \frac{w}{t}$$

$$\text{Efficiency} = \frac{\text{Mechanical advantage}}{\text{Velocity ratio}} \qquad \eta = \frac{MA}{VR}$$

$$\text{Efficiency} = \frac{\text{Output}}{\text{Input}} = \frac{\text{Output}}{\text{Output} + \text{Losses}} = \frac{\text{Input} - \text{Losses}}{\text{Input}}$$

$$\text{Mechanical advantage MA} = \frac{\text{Load force}}{\text{Effort force}}$$

$$\text{Velocity ratio VR} = \frac{\text{Distance moved by effort}}{\text{Distance moved by load}}$$

$$\text{Power transmission} = \frac{N_1}{N_2} = \frac{D_2}{D_1} = \frac{T_2}{T_1}$$

- N_1 is the speed of the first pulley or gear, r/min
- N_2 is the speed of the second pulley or gear, r/min
- D_1 is the diameter of the first pulley or gear, m
- D_2 is the diameter of the second pulley or gear, m
- T_1 is the torque provided to, or by, the first pulley or gear, Nm
- T_2 is the torque provided to, or by, the second pulley or gear, Nm

5.8   Exercises

1.   A sling is marked as having a safe working load of 2000 kg. What weight will it support safely?
2.   A machine requires a force of 500 N to lift it. What is its mass?
3.   A masonry drill has an effective surface area of 50 mm^2 and, for best rate of penetration, must be operated at a pressure of 4 N/mm^2. What force should be applied to the drill?
4.   An apprentice weighs 480 N. What pressure will be exerted on a plasterboard ceiling if he stands between the joists
     (a)   on one foot which has an area of contact with the ceiling of 8000 mm^2
     (b)   on a rigid plank 300 mm wide and 2 m long?
5.   A turning moment of 18 Nm is required to tighten a nut on a busbar clamp. What maximum force must be applied to a spanner of effective length 0.27 m if the nut is not to be overtightened?
6.   An electrician exerts a turning force of 50 N on each of two handles of a set of stocks and dies. The effective length of each handle is 0.2 m. What total turning moment is applied to the dies?
7.   How much energy must be expended to raise a bundle of conduit weighing 600 N from a floor to a scaffolding 7 m above it? If the task takes two minutes to complete, what average power is used?
8.   A motor has timbers bolted to it for protection and requires a horizontal force of 450 N to move it over a level floor. How much work is done in moving it 50 m?
9.   The petrol engine of a builder's hoist is to be replaced by an electric motor. What is the power rating of the motor if it must be capable of lifting 2400 N through 32 m in 24 s, the hoist gear being 80% efficient?
10. The pivot of a hydraulic pump is at one end of the handle, which is 1.2 m long, and is 0.2 m from the attachment to the pump rod. The pump piston has an effective surface area of 200 mm^2. If the handle of the pump is pushed

down by a force of 80 N, calculate (a) the force on the pump rod and (b) the pressure on the pump piston.

11. What must be the minimum weight of a man who is able to lift a load of 5600 N using a crowbar 2 m long, with the fulcrum 0.2 m from the load end, and what is the mechanical advantage?

12. A block and tackle of eight pulleys is to be used to lift a cubicle switch panel weighing 2400 N. The lifting system is 80% efficient. Calculate the effort required.

13. A screw jack has a velocity ratio of 112 and an efficiency of 50%. What force must be applied to it to lift a drum of cable weighing 8960 N?

14. A load pulley is 0.5 m in diameter, turns at 3.33 r/s, and has a torque of 800 Nm. If the motor speed is 12 r/s, calculate the size of the motor pulley and the torque it provides.

15. A load shaft is required to turn at 12.08 r/s and to provide a torque of 1000 Nm. If the pulley on the 24.17 r/s motor is 250 mm in diameter, calculate the diameter of the load pulley and the torque provided by the motor.

16. A motor providing a torque of 100 Nm at 2900 r/min is fitted with a gearwheel having 20 teeth. A drive is to be provided to a shaft having a gearwheel in mesh with that on the motor. If the shaft must provide a torque of 650 Nm, how many teeth has the driven gear, and what is its speed?

17. Two horizontal cables are attached to a pole. The first exerts a force of 2000 N, and the second a force of 3200 N making an angle of 120° with the first. Find the resultant pull on the pole and its direction with respect to the first cable.

18. Two runs of cable exert forces of 6000 and 8000 N, respectively, on an overhead-line tower to which they are attached at right angles. What must be the direction of a stay, and what will be the horizontal force it supports, if there is to be no resultant horizontal force on the tower?

19. A set of pulley blocks, having four pulleys in the top block and three pulleys in the lower block, is fixed to a ceiling beam in a workshop. It is to be used to lift a motor weighing 2520 N from its bedplate. The efficiency of the tackle is 60%. Calculate the pull required on the free end of the rope to raise the motor. Make a diagrammatic sketch of the tackle in use.

20. A motor armature weighing 2000 N is freely suspended from a crane hook by means of a double sling with 1 m chains. The motor shaft is horizontal, and the slings are attached to the motor shaft 1 m apart.
(a) Draw a diagram showing the arrangement.
(b) Determine the tension in the sling.
(c) What would be the tension if the chains were 0.835 m instead of 1 m long?

5.9 Multiple-choice exercises

5M1 The SI unit of mass is the
(a) newton
(b) weight

(c) kilogram
(d) force

5M2 A man exerts a force due to gravity of 750 N when he stands on one foot which is subject to a pressure of 30 kN/m^2. What is the area of his foot?
(a) 250 cm^2
(b) 40 cm^2
(c) 25 cm^2
(d) 2.5 cm^2

5M3 The density of a piece of material can be defined as
(a) how heavy it is
(b) how thick it is
(c) if it floats in water
(d) the mass of one cubic metre of it

5M4 The unit of torque, the turning moment of a force, is the
(a) newton meter (Nm)
(b) angular twist in degrees
(c) force on a lever
(d) metre newton (mN)

5M5 Work (or energy) can be defined as
(a) something hard
(b) force exerted times distance moved
(c) power in watts
(d) the force required to turn a spanner

5M6 The unit of energy is the
(a) watt
(b) kilogram
(c) newton metre
(d) joule

5M7 Efficiency can be defined as
(a) $\dfrac{\text{input}}{\text{output}}$
(b) $\dfrac{\text{output} - \text{losses}}{\text{output}}$
(c) $\dfrac{\text{output}}{\text{output} + \text{losses}}$
(d) $\dfrac{\text{output} + \text{losses}}{\text{losses}}$

5M8 An electric motor drives a pump which lifts 12 litres (118 N) of water to a height of 8 m. If the combined efficiency of the motor and the pump is 35%, the motor input power will be
(a) 330 W
(b) 2.70 kW

(c) 274 W

(d) 944 W

5M9 A class 1 lever is 2.4 m long and is used to lift a load weighing 2000 N at a distance of 0.4 m from the fulcrum. The effort required will be

(a) 400 N

(b) 333 N

(c) 12 000 N

(d) 10 000 N

5M10 A jack used to lift heavy loads is

(a) the name of an extremely strong man

(b) a form of class 2 lever

(c) a complicated form of block and tackle

(d) an inclined plane wound round a central column to form a screw thread

5M11 A block and tackle has a total of six pulleys and is 75% efficient overall. If the maximum possible effort on the operating rope is 200 N, it will be possible to lift a load with a maximum weight of:

(a) 1200 N

(b) 1600 N

(c) 25 N

(d) 900 N

5M12 A motor with a speed of 18 r/s in a clockwise direction is fitted with a vee belt pulley of diameter 20 cm. If the pulley is connected to a drive pulley of diameter 30 cm, the speed and direction of the driven pulley will be:

(a) 27 r/s anticlockwise

(b) 12 r/s clockwise

(c) 27 r/s clockwise

(d) 12 r/s anticlockwise

5M13 Two gears have 24 teeth and 120 teeth, respectively, and are meshed together. If the output is taken from the 120-tooth gear, which provides a torque of 15 Nm at a speed of 200 r/min, the speed and torque of the 24-tooth driven gear, assuming no losses, will be

(a) 1000 rev/min, 3 Nm

(b) 500 r/min, 75 Nm

(c) 40 r/min, 75 Nm

(d) 1000 r/min, 6 Nm

5M14 A vector quantity is one

(a) that has magnitude but no direction

(b) whose value is known exactly

(c) that has a magnitude and a direction

(d) that measures quantities such as length, mass and power

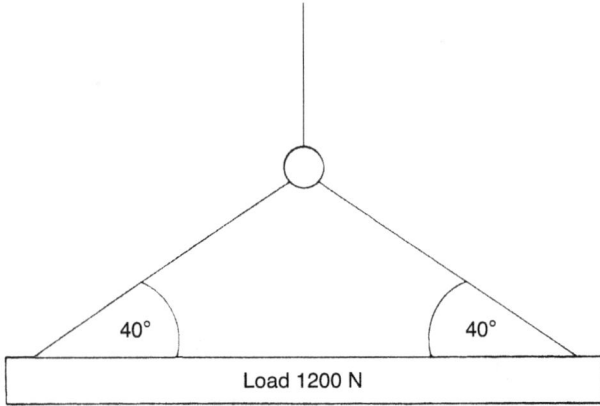

Figure 5.11   Diagram for Exercise 5M16

5M15 If forces are equal and opposite so that no movement results, they are said to
be
(a) fixed
(b) in equilibrium
(c) a resultant
(d) vectors

5M16 A load weighs 1200 N and is suspended from a single chain, the end of
which is connected to two others as shown in Figure 5.11. The tension in
each of the two chains will be
(a) 1200 N
(b) 905 N
(c) 933 N
(d) 1540 N

Chapter 6

Electrical power and energy

6.1 Units of electrical power and energy

We found in Chapter 3 that a PD of one volt exists between two points if one joule of energy is expended in moving one coulomb of electricity between them. Thus,

Energy = Charge × Volts $\qquad$ $w = QU$

Since the charge is Current × Time $\quad$ $Q = It$

Energy = Volts × Current × Time $\quad$ $w = UIt$

In Chapters 4 and 5, we see that power is the rate of using energy. The mechanical unit of power is the watt (W). This is also the unit for electrical power. There is no reason for different units of power to exist because the work done at a given rate can be the same whether provided by an electric motor or by a mechanical method such as a diesel engine.

The watt is defined as a rate of doing work of one joule per second. Therefore,

$$P = \frac{\text{Energy}}{\text{Time}} = \frac{UIt}{t} = UI$$

In other words, the power in watts at any instant in a given circuit is the instantaneous circuit PD in volts multiplied by instantaneous current circuit in amperes, or

$P = UI$ for DC circuits

It is often useful to be able to express the power (P) in a circuit in terms of its resistance, together with either current or PD. This method is used to calculate the power wasted by copper losses in the windings of a transformer.

$P = UI$, but $U = IR$, so $P = I(IR) = I^2R$

Also

$$P = UI, \text{ but } I = \frac{U}{R}, \text{ so } P = \frac{UU}{R} = \frac{U^2}{R}$$

Example 6.1
A 100 Ω resistor is connected to a 10 V DC supply. How much power is dissipated?

$$P = \frac{U^2}{R} = \frac{10^2}{100} = 1 \text{ W}$$

or

$$I = \frac{U}{R} = \frac{10}{100} = 0.1 \text{ A}$$

$$P = UI = 10 \times 0.1 = 1 \text{ W}$$

or

$$P = I^2 R = 0.1^2 \times 100 \text{ W} = 1 \text{ W}$$

Example 6.2
What is the hot resistance of a 230 V, 100 W lamp?

$$P = \frac{U^2}{R}, \text{ so } R = \frac{U^2}{P} = \frac{230^2}{100} = 529 \text{ Ω}$$

Since $P = \frac{w}{t}$, it follows that the joule, the unit of energy, is the watt-second.

Example 6.3
How much energy is supplied to a 100 Ω resistor, which is connected to a 150 V supply for 1 h?

$$P = \frac{U^2}{R} = \frac{150^2}{100} = 225 \text{ W}$$

Convert hour to seconds (the SI unit for time) 1 h = 3600 s

$$w = Pt = 225 \times 3600 = 810,000 \text{ J or } 810 \text{ kJ}$$

Note that the joule is too small for the measurement of the amounts of energy in common use, so the kilowatt hour (kWh) is the unit for practical purposes. The kilowatt hour is the energy used when a power of 1 kW (1000 W) is used for 1 h (3600 s) and is often referred to as the 'unit' of energy. The kilowatt hour (kWh) is also used as the standard unit for the measurement of gas supplies.

Example 6.4

A DC motor takes 15 A from a 230 V supply, and is used for 40 min. What will this cost, if the tariff is 25 p/kWh?

Although a second is the SI unit for time, we need to express time in hours in this case because that is how energy use is measured for real-life electricity charges.

To convert 40 min to hours $\frac{40}{60} = 0.67$ h

$$\text{Energy} = \frac{U \times I}{1000} \times \text{hours} = \frac{230 \times 15}{1000} \times 0.67 = 2.31 \text{ kWh}$$

Energy charge $= 2.31 \times 25$ pence $= 57.79$ p

Example 6.5

An electric heater consumes 2.7 MJ when connected to its 230 V supply for 30 min. Find the power rating of the heater and the current taken from the supply.

$$\text{Power} = \frac{\text{Energy}}{\text{Time}} \quad P = \frac{w}{t}$$

Convert minutes to seconds: $30 \times 60 = 1800$ s

$$P = \frac{2.7 \times 10^6}{1800} = 1500 \text{ W or } 1.5 \text{ kW}$$

$$P = UI \text{ transpose to make } I \text{ the subject } I = \frac{P}{U} = \frac{1500}{230} = 6.52 \text{ A}$$

Example 6.6

A house has the following circuit loads:

- Lighting (all LED):
  o Two 50 W lamps
  o Six 100 W lamps
  o Ten 60 W lamps
- Two-ring final circuits
- One 2 kW heater
- One 9 kW cooker
- One 8.5 kW electric shower

(a) What is the current in each circuit if the supply is at 230 V?
(b) What size protective device should be used for each circuit?
(c) What is the load and protective device rating for the whole installation?

Lighting

Total load comprises

$$2 \times 15 \text{ W} = 30 \text{ W}$$
$$6 \times 7.5 \text{ W} = 45 \text{ W}$$
$$10 \times 11 \text{ W} = 110 \text{ W}$$
$$\text{Total} = 185 \text{ W}$$

$$I = \frac{P}{U} = \frac{185}{230} = 0.8 \text{ A}$$

A suitable protective device rating would be 1 A. However, the standard rating for a lighting circuit is 6 A.

Note that it is not good practice to have 18 lighting points all on the same circuit because there will be no light in the building if the circuit fails.

Ring final circuits

According to BS7671 and the IET On-Site Guide a Type A Ring Final Circuit feeding 13 A socket outlets should be protected by a 32 A device. The calculations for this appear in Chapter 15.

2 kW heater

$$I = \frac{P}{U} = \frac{2000}{230} = 8.7 \text{ A}$$

A suitable protective device rating would be 10 A.

9 kW cooker

$$I = \frac{P}{U} = \frac{9000}{230} = 39.13 \text{ A}$$

A suitable protective device rating would be 40 A.

8.5 kW electric shower

$$I = \frac{P}{U} = \frac{8500}{230} = 36.96 \text{ A}$$

A suitable protective device rating would be 40 A.

Total installation load

Lights	0.8 A
Ring final circuits	64 A
Heater	8.7 A
Cooker	39.13 A
Shower	36.96 A
Total	149.59 A

The usual rating for the protective device protecting a house is 100 A. This rating is derived from the application of diversity to the load current. Diversity is based on the assumption that not all the circuits will run at full capacity at the same time (see Volume 2).

6.2   Electromechanical conversions

In SI units, there is no difference between mechanical and electrical quantities. The watt is the unit of power in all systems, and is always defined as the rate of doing the work of one joule per second.

Similarly, the joule is the unit of energy in both mechanical and electrical systems.

- In mechanical applications, the joule is defined as the work that is done when a force of one newton is moved through 1 m.
- For electrical applications, the joule can be defined as the work done when one coulomb of electricity is moved through a potential difference of one volt.

Since the same units are used for both systems, conversions from mechanical to electrical forms, and vice versa, rely on the following relationship:

$$\text{Joules} = \text{coulombs} \times \text{volts} = \text{newtons} \times \text{metres} \quad w = QU = Fd$$

Since

$$\text{Coulombs} = \text{amperes} \times \text{seconds} \quad Q = It$$

$$\text{Watts} = \frac{\text{joules}}{\text{seconds}} = \text{amperes} \times \text{volts} = \frac{\text{newtons} \times \text{metres}}{\text{seconds}}$$

$$P = \frac{w}{T} \quad UI = \frac{Fd}{t}$$

Example 6.7
Calculate the power rating of a lift that can raise a mass of 800 kg through a height of 5 m in 12 s.

$$w = mgd \quad w = 800 \times 9.81 \times 5 = 39{,}240 \text{ J}$$

$$P = \frac{w}{t} = \frac{39{,}240}{12} = 3270 \text{ W or } 3.27 \text{ kW}$$

Example 6.8
Calculate the output and input powers of a motor driving the lift of Example 6.7 if the lift gear is 80% efficient and the motor is 90% efficient.

$$\text{Efficiency} = \frac{\text{Output}}{\text{Input}}$$

$$\text{Input to the lift (which is also the output of the motor)} = \frac{\text{Output}}{\text{Efficiency}}$$

$$\text{Output of motor} = 4 \times \frac{100}{80} = 5 \text{ kW}$$

$$\text{Motor input} = \frac{\text{Output}}{\text{Efficiency}} = 5 \times \frac{100}{90} = 5.56 \text{ kW}$$

Example 6.9
If the motor of Examples 6.7 and 6.8 is fed from a 220 V DC supply, what current will it take during the lift?

$$P = UI, \text{ so } I = \frac{P}{U} = \frac{5560}{220} = 25.3 \text{ A}$$

6.3   Exercises

1. An electric fire takes a current of 6.25 A from a 230 V supply. What power does it dissipate?
2. How much power will be dissipated in a 3 Ω resistor when it is connected across a 12-V battery?
3. A hall is lit by 20 lamps, each of which takes 0.5 A when connected to a 230 V supply. Calculate the total power supply to the lamps.
4. What current will a 750 W fire element take from a 230 V supply?
5. How much current will a 10 Ω resistor be carrying if the power lost in the resistor is 1 kW?

6.   A 3 kW heating element has a resistance of 30 Ω. For what supply voltage was the heater designed?

7.   An electric water heater must provide 4 MJ of energy to heat its contents in 50 min. What is the power rating of the heating element?

8.   An underfloor heating system has a loading of 20 kW. How much energy will it consume in 10 h?

9.   A motor with an output of 8 kW and an efficiency of 80% runs for 8 h. Calculate the cost of electricity costs 15 pence per unit.

10. A generator in a motor car charges the battery at 20 A at 16 V. The generator efficiency is 43%. What power is absorbed in driving it?

11. It costs 40 pence to run a motor for 2 h, the tariff being 22 pence per unit. If the motor is 90% efficient, what is its output?

12. An electric water heater provides 4.8 MJ in 40 min. What is the rating of the element and how much current does it take from a 230 V supply?

13. A domestic load comprises:
     (a) 12 × 7 W lamps (the lamps are fed from a single lighting circuit)
     (b) 8 × 11.5 W lamps
     (c) 2 × 2 kW radiators
     (d) 1 × 3 kW shower
     The supply is at 230 V. What is the current in each circuit? Suggest a suitable protective rating for each circuit.

14. A small flat has the following loads:
     (a) 3 × 11.5 W lamps in use 5 h each day
     (b) 1 × 2 kW heater in use 4 h each day
     (c) 1 × 9 kW shower in use 0.5 h each day
     (d) 1 × 4 kW cooker in use 2 h each day
     The tariff for the supply of electricity is a fixed charge of 18 pence per week, plus an energy charge of 5 pence per unit. Find the total electricity cost for one week.

15. A lift motor takes a current of 20 A from a 300 V DC supply. The lift gearing is 50% efficient. If the lift is able to raise a load of 1440 N through a height of 50 m in 30 s, what is the efficiency of the motor?

16. Calculate the output of a motor with an input of 24 kW and an efficiency of 90%.

17. A motor with an output of 10 kW and an efficiency of 85% drives a drilling machine, which is 60% efficient. What is the output of the machine, and what is the input power?

6.4   Multiple-choice exercises

6M1 The unit of electrical power is the
     (a) volt
     (b) ampere
     (c) watt
     (d) horsepower

6M2 The power dissipated in an 800 resistor connected to a 230 V supply is
  (a) 66.1 W
  (b) 0.3 A
  (c) 7.2 W
  (d) 192 W

6M3 An electric lamp is rated at 60 W. The current it draws from a 230 V supply is
  (a) 4 A
  (b) 14.4 A
  (c) 6.94 μA
  (d) 0.261 A

6M4 If the 60 W lamp of question 5M3 is left on for 8 h and electrical energy costs 7 pence per unit, the cost will be
  (a) 3.36 p
  (b) £3.36
  (c) 5.25 p
  (d) 6.86 p

6M5 The energy used by the 60 W lamp of exercise 5M4 in 8 h is
  (a) 480 J
  (b) 172.8 kJ
  (c) 28.8 kJ
  (d) 60 W

6M6 A resistor of value 33 Ω is connected to a 100 V supply through a cable with a total resistance of 0.3 Ω. The thermal energy in the cable is
  (a) 300 W
  (b) 3.63 W
  (c) 2.71 W
  (d) 33.3 kW

6M7 Resistors are connected in a series–parallel group as shown in the figure below. The power dissipated in the 5 Ω resistor is
  (a) 2.56 W
  (b) 45 W
  (c) 80 W
  (d) 4 A

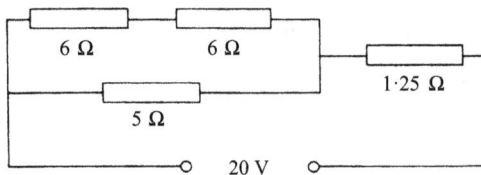

6M8   The SI unit of force is the
   (a) pound
   (b) horsepower
   (c) joule
   (d) newton

6M9   The number of newtons in 1 kg is
   (a) 9.81
   (b) 33,000
   (c) 453.6
   (d) 3,600,000

6M10 If 120 J of energy is used to move a force of 10 N, the distance moved is
   (a) 1.2 m
   (b) 12 m
   (c) 120 cm
   (d) 0.0

Chapter 7

Heat

7.1 Heat

In early childhood, we become familiar with the sensations of cold and warmth and are able to distinguish between them. We learn to estimate degrees of hot and cold. This is known as the temperature.

Heat is a form of energy.

- Heat added to a body makes it hotter
- Heat taken away from a body makes it colder.

It is possible to increase the heat energy contained in a piece of metal and, as a result, increase its temperature, by cutting, bending or hammering. It will also become hotter if work is done against friction, or if a fire is lit beneath the metal. This is because it absorbs part of the energy made available to it. So, when heat energy is produced, energy in some other form is expended.

Most of the losses of energy that occur in machines appear as heat, which means that energy is lost to the process concerned. Remember, energy can be changed but not destroyed.

7.2 Temperature

Temperature is a measure of the degree of hotness or coldness of a body. We are often concerned with the accurate measurement of temperature. Temperature measurement can be carried out using a thermometer. Many thermometers rely for their operation on the property of certain materials to expand as their temperature increases.

Glass-bulb thermometers consist of a mercury, or alcohol-filled, bulb attached to the bottom of a fine glass tube. As the heat increases, the liquid expands and is forced up the tube. An adjacent scale, indicates the temperature according to how far the liquid rises.

Another thermometer type consists of a long metal strip in the form of a coil. One end is securely fixed, and movement of the other end as the strip expands moves a pointer over a scale.

Electronic thermometers are now common and rely on one of two operating principles.

Thermocouple

A thermocouple consists of a junction between two different metals that experience a very small potential difference (Figure 7.1). The value of the potential difference depends on the temperature of the junction. This potential difference is measured by the instrument and converted to temperature, which is displayed digitally.

Thermistor

The second type uses a thermistor (a resistor whose resistance changes with temperature) enclosed in a probe (Figure 7.2). The probe is placed at the point where the temperature measurement is required. The thermometer measures the resistance of the thermistor and calculates its temperature.

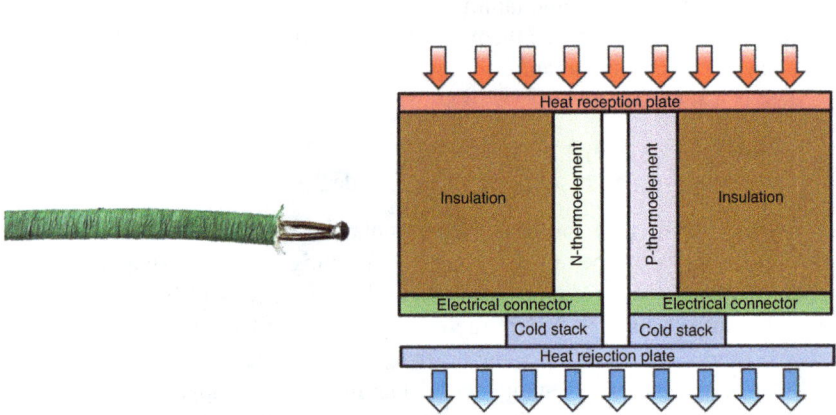

Figure 7.1   Example of thermocouple (left), working principle of the thermocouple (right). Image credits: (a) Harke, via Wikimedia Commons and (b) Unknown, via Wikimedia Commons, CC BY-SA 4.0.

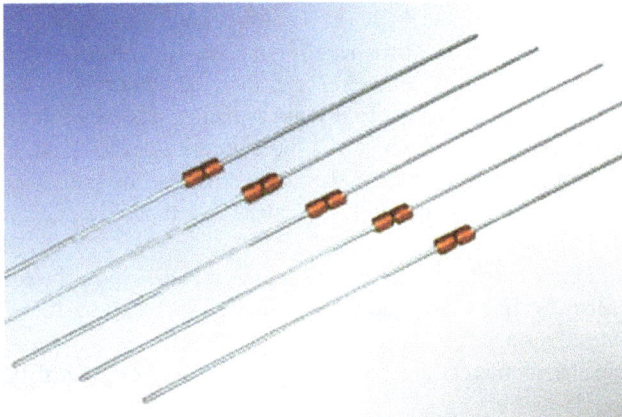

Figure 7.2   A thermistor

As well as being more accurate and more easily read than the other types of thermometer, the electronic type has the advantage that the protected probe can often be put in positions not accessible to normal thermometers. For example, the probe may be driven into frozen food, lowered deep into liquid and so on.

Temperature scales

Before any scale of measurement can be decided, two fixed points are necessary. For instance, in length measurement, the distance between two marks on a metal bar can indicate a standard length. For temperature scales:

- The upper fixed point is the temperature of steam from boiling pure water
- The lower fixed point is the temperature at which ice just melts, both measured at normal atmospheric pressure.

Between these fixed points, the scale may be subdivided into any suitable number of parts.

- The Celsius scale takes the lower fixed point as zero degrees Celsius (written 0 °C) and the upper point as 100 °C
- The zero degrees Kelvin scale is the lowest possible temperature (called 'absolute zero') written as 0 K. In the Celsius scale, absolute zero is −273 °C

A temperature change of 1 kelvin is the same as one degree Celsius, so if we say that 0 K = −273 °C, it follows that 0 °C = 273 K. Thus,

- Any temperature expressed in degrees Celsius can be given in kelvins by the addition of 273
- Any temperature given in kelvins can be converted to degrees Celsius by subtraction of 273

Example 7.1

1. Convert to kelvin
   (a) 20 °C
   (b) 230 °C
   (c) −40 °C

2. Convert to degrees Celsius
   (a) 265 K
   (b) 300 K
   (c) 1200 K

1. (a) 20 °C = (20 + 273) kelvin = 293 K
   (b) 230 °C = (230 + 273) kelvin = 503 K
   (c) −40 °C = (−40 + 273) kelvin = 233 K

2. (a) 265 K = (265 − 273) degrees Celsius = −8 °C
    (b) 300 K = (300 − 273) degrees Celsius = 27 °C
    (c) 1200 K = (1200 − 273) degrees Celsius = 927 °C

Strictly speaking, the Kelvin scale is the SI unit for temperature. However, the Celsius scale has more convenient numbers for everyday temperatures, and its widespread use is likely to continue.

7.3   Heat units

Because heat is a form of energy, its unit of heat is the joule.

It can be shown experimentally, but not proved mathematically, that it takes 4187 J of heat energy to heat one litre of water (mass 1 kg) through 1 °C.

This figure 4187 J is called the specific heat of water. Different substances have different specific heats, but in every case the specific heat is the heat energy needed to raise the temperature of 1 kg of the substance by 1 K.

The temperature of a body depends not only on the heat contained in the body, but also on its mass and its capacity to absorb heat. For instance:

* A small bowl of very hot water will be at a higher temperature than a bath full of cold water, but may contain less energy owing to its smaller mass.
* 1 kg of brass will increase in temperature by 1 K for the addition of 376.8 J, whereas the same mass of water requires 4187 J for this temperature change.

It is clear, then, the quantity of energy which must be added to a body to raise its temperature (or which must be removed from a body to lower its temperature) depends on its:

* Mass
* Specific heat
* The temperature change involved

Heat energy for temperature change = Mass × Temperature change

K × Specific heat = J/kg/K

Example 7.2
An immersion heater is required to raise the temperature of 50 litres of water from 10 °C to 85 °C. If no heat is lost, find the energy required.

Energy required = Mass × Temperature change × Specific heat

Table 7.1   Specific heats

Substance	Specific heat (J/kg/K)
Water	4187
Air	1010
Aluminium	915
Iron	497
Copper	397
Brass	376
Lead	129

From Table 7.1, the specific heat of water is 4187 J/kg/K. One litre of water has a mass of 1 kg.

Energy required $= 50 \times (85 - 10) \times 4187 = 50 \times 75 \times 4187$

$$= 15{,}700{,}000 \text{ J or } 15.7 \text{ MJ}$$

In practice, heat will be lost through the container during heating, so the efficiency of the system must be taken into account.

Example 7.3
A tank containing 150 litres of water is to be heated from 15 °C to 75 °C. If 25% of the heat provided is lost through the tank, how much energy must be supplied?
From Table 7.1, the specific heat of water is 4187 J/kg/K.
NOTE: If 25% of the heat is lost, 75% is retained, which means that the system is 75% efficient.

Energy required at 100% efficiency $= 150 \times (75 - 15) \times 4187 = 37{,}683{,}000 \text{ J}$

The energy required at 75% efficiency will be greater.

$$= 150 \times (75 - 15) \times 4187 \times \frac{100}{75}$$

$$= 150 \times 60 \times 4187 \times \frac{100}{75} = 50{,}240{,}000 \text{ J or } 50.24 \text{ MJ}$$

Example 7.4
Two kilograms of iron are placed in a furnace at an initial temperature of 15 °C, and heat energy of 3.976 MJ is provided to the furnace. If one-quarter of the heat provided is received by the iron, what will be its final temperature?
From Table 7.1, the specific heat of iron is 497 J/kg/K.

If the total heat provided is 3.976 MJ, and one-quarter of this reaches the iron, the heat absorbed by the iron is

$$\frac{3.976\text{ MJ}}{4}, \text{ or } 0.994\text{ MJ}$$

Heat provided = Mass × Temperature change × Specific heat

$$\text{Temperature change} = \frac{\text{Heat provided}}{\text{Mass} \times \text{Specific heat}} = \frac{0.994 \times 10^6}{2 \times 497} = 1000\text{ K}$$

If the initial temperature is 15 °C and increases by 1000 K, the final temperature will be 1015 °C.

7.4   Heating time and power

We have already seen that a rate of doing work of one joule per second is a power of one watt.

$$w = Pt$$

It is common for the kilowatt hour (kWh) to be used as a larger and more convenient unit than the joule. A kilowatt hour is the energy used when a power of 1000 W is applied for 1 h.

Note: There are 3600 s in 1 h.

$$1\text{ kWh} = 1000\text{ W} \times 3600 = 3{,}600{,}000\text{ W/s or J/kg K}$$

$$1\text{ kWh} = 3.6\text{ MJ}$$

The rating of electric heaters is usually in watts or kilowatts, and the time relationship of this power with heating energy enables us to calculate the time taken for a heater of a certain rating to provide a particular quantity of heat.

The rating necessary to provide a given quantity of heat in a given time can also be found.

These calculations are illustrated in the following examples. Although electric heating is probably the most efficient of the heating methods, some heat may be lost, and this must be taken into account.

Example 7.5
How long will a 3 kW immersion heater take to raise the temperature of 30 litres of water from 10 °C to 85 °C? Assume that the process is 90% efficient.

Table 7.1 gives the specific temp of water as 4187 J/kg/K.

$$\text{Heat required} = 30 \times (85 - 10) \times 4187 \times \frac{100}{90} = 10{,}470{,}000 \text{ J or } 10.47 \text{ MJ}$$

A 3 kW heater gives 3000 W (or 3000 J/s), therefore

$$\text{Time required} = \frac{10{,}470{,}000}{3000} = 3490 \text{ s or } \frac{3490}{60} = 58.17 \text{ min}$$

Example 7.6

A jointer's pot contains 25 kg of lead and is heated by a gas flame providing a power of 2 kW to the pot. If the initial temperature of the lead is 15 °C, how hot will it be after 5 min?

From Table 7.1, the specific heat of lead is 129 J/kg/K.

We can assume 100% efficiency, since 2 kW is provided to the pot. The heat lost will be several times this figure for a heater of this type.

Heat provided = 2 kW for 5min (300 s) = 2000×300 = 600,000 J

Heat used = Mass × Temperature × Specific heat

Since heat provided = Heat used

$600{,}000 = 5 \times$ temperature change $\times\ 129$

Transpose to make the temperature change the subject

$$\text{Temperature change} = \frac{600{,}000}{25 \times 129} = 186 \text{ K}$$

Since the initial temperature is 15 °C and the increase is 186 K, the final temperature is $15 + 186 = 201$ °C.

Example 7.7

A 3 kW instantaneous water heater is assumed to be 100% efficient. What will be the increase in water temperature from the inlet to the outlet if the rate of flow is 1 litre/min?

Table 7.1 gives the specific temp of water as 4187 J/kg/K.

Since 3 kW = 3000 J/s, heat provided in 1min = 3000×60 = 180,000 J

Heat provided = Mass × Temperature change × Specific heat

$180{,}000 = 1 \times$ Temperature change $\times\ 4187$ J

Transpose to make temperature the change the subject

$$\text{Temperature change} = \frac{180{,}000}{1 \times 4187} = 43 \text{ K or } 43\,°\text{C}$$

7.5   Heat transmission

In Chapter 5, we said that some of the heat produced by a machine is 'lost' to the atmosphere. This indicates clearly that heat must be able to move from the point at which it is generated. Transmission of heat is an essential part of many engineering processes. For instance, heat must be transferred from the tip of a soldering iron to the solder and to the surfaces to be joined. Again, heat is given out from the element of an electric fire. If this didn't happen, the continuous input of energy by means of an electric current would increase the temperature of the element until it melted.

The amount of heat transmitted depends on the difference in temperature between the heat source and the places into which the heat is transferred. When an electric fire is first switched on, very little heat is transmitted because the element temperature is low.

As time goes by, the electrical energy fed into the element appears as heat, raising the temperature and increasing the amount of transmitted heat. In due course, the element becomes hot enough to transmit the same amount of heat energy as it receives. In this condition of heat balance, the temperature of the element remains constant. This heat balance applies to all devices in which heat is generated. The final steady temperature depends on the energy input and the means by which heat can be transmitted.

In practice, heat is transmitted by three separate processes, which can occur individually or in combination. These processes are called

- Conduction
- Convection
- Radiation

Conduction

If one end of a metal bar is heated, the other end also becomes hot. Heat is conducted along the length of the bar from the high-temperature end to the low-temperature end as the heat energy tries to distribute itself evenly throughout the bar. If we attempt a similar experiment with wood, however, the heated end will burn without the colder end experiencing a significant rise in temperature. Metal, therefore, is said to be a good conductor of heat. Wood is a poor conductor.

Most liquids and gases are poor conductors of heat. Because of this, materials which trap air in pockets, such as felt, glass fibre and cork are used to prevent the conduction of heat, and are called heat insulators. An application of such materials, often called lagging, is used to prevent heat escape from ovens, hot-water tanks and the like. The same sort of lagging may be used to prevent heat *entering* a refrigerator.

Convection

Transmission of heat by convection takes place in liquids and gases. If a given volume of a liquid or a gas is heated, it expands (if it is free to do so). As a result, the same volume weighs less than that of the unheated fluid. The density has decreased. The heated fluid will rise, its place being taken by cool fluid, which will also rise when

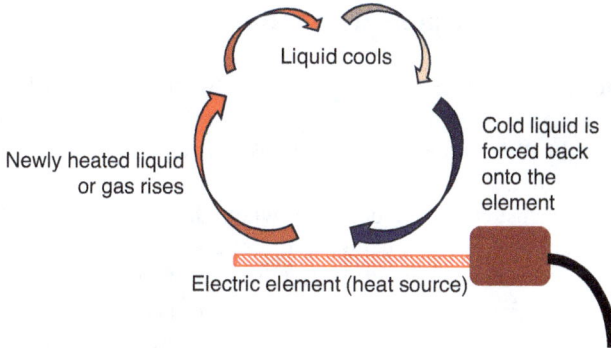

Figure 7.3   Simplified diagram showing convection

heated, so that a steady upward movement of warm fluid results. When the fluid cools is descends back to the heat source to be re-heated and so the cycle continues (see Figure 7.3).

This principle is widely used in some types of air heaters called convectors, which draw in cool air at the bottom and expel hot air from the top.

The same sort of process is used in some types of central-heating system, where water circulates through the radiators solely as a result of convection currents. (In large installations, or those using small-bore piping, the natural circulation is assisted by a pump.) The contents of a kettle also circulate owing to convection.

Water is, however, a very bad conductor of heat, and the water below the level of a heater element remains cold and is not displaced by convection currents.

Radiation

Most of the energy reaching the Earth does so in the form of heat radiated from the Sun. Unlike conduction and convection, which can occur only in material substances, radiation of heat from one body to another does not require any connecting medium between the bodies.

Heat is radiated freely through space in much the same way as light. All bodies constantly radiate thermal energy. The rate of radiation depends on the temperature of the body and its surroundings. A cold body surrounded by warm objects will radiate less heat than it receives, and will become warmer.

Radiant heat behaves in a very similar manner to light and is reflected from bright surfaces. A highly polished reflector is fitted to a fire with a rod-type element, and this reflector beams the heat in a particular direction. Fires of this sort are often called radiators, although much of their energy is given off in the form of convection, and a little by conduction.

A body with a dark, matt surface tends to absorb heat instead of reflecting it. If the reflector of a radiator were to be painted black, it would absorb heat quickly and become excessively hot. Bright surfaces do not emit (as opposed to reflect) radiation as readily as dull ones. For this reason, appliances such as metal electric kettles are often chromium plated to reduce heat losses and improve efficiency.

7.6   Change of dimensions with temperature

The dimensions of most materials increase slightly when their temperature increases. For instance, overhead lines sag more in the summer than in the winter, and a long straight PVC conduit run may buckle in very hot weather if expansion couplers are not fitted.

Similarly, dimensions often decrease when the temperature decreases. An overhead cable erected in hot weather and strained tightly will contract in cold weather. The extra stress may result in stretching or even breaking.

Although expansion with increasing temperature is often a nuisance, it can be used to control temperature. Some metals expand more than others when heated through the same temperature range. If two strips of two metals with different rates of expansion are riveted together to form a bimetal strip, it will bend when heated. If a set of contacts is operated by the strip as it bends, the device can be made to control temperature and is called a thermostat (Figure 7.4).

Another type of thermostat based on the same principle is called a rod-type thermostat (Figure 7.5). A rod of material, selected for its small increase in length when heated, is mounted within a tube of brass. The rod and the tube are welded together at one end. Changes in the temperature of the device result in differing changes in length of the rod and the tube, which can be used to operate a switch. The slow break resulting from the slow rate of differential expansion will give rise to arcing at the contacts of a directly operated switch. Permanent-magnet systems and flexed springs are often used to give a quick make-and-break action, which will reduce damage from arcing.

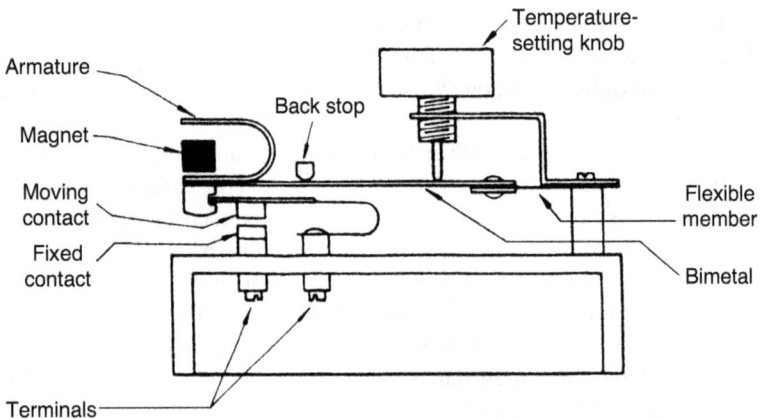

Figure 7.4   Air thermostat

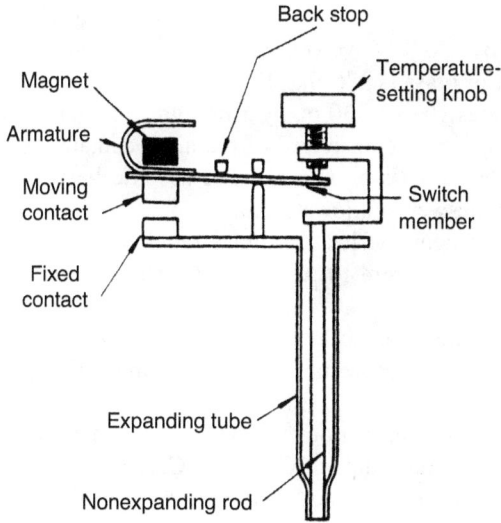

Figure 7.5   Rod-type immersion thermostat

7.7   Summary of formulae for Chapter 7

Temperature in kelvin = Temperature in degrees Celsius + 273

Temperature in degrees Celsius = Temperature in kelvin − 273

Heat energy for temperature change J = Mass, kg × Temperature change
                                              × Specific heat

Energy, J = Power, W × Time, s $w = Pt$

7.8   Exercises

1. Express the following temperature in kelvin:
   (a) 60 °C
   (b) −75 °C
   (c) 1000 °C

2. Express the following temperatures in degrees Celsius:
   (a) 320 K
   (b) 1500 K
   (c) 240 K

3. A small storage heater contains 8 l of water at a temperature of 10 °C. How much heat energy must be provided to raise the water temperature to 90 °C? The specific heat of water is 4187 J/kg/K.

4. How much heat energy must be supplied to 20 kg of brass to increase its temperature by 500 K? Give your answer in joules and in kilowatt hours. The specific heat of brass is 376 J/kg/K.

5. A hall measures 10 m by 30 m, by 5 m high. How much heat energy will be required to raise the temperature of the air it contains from 5 °C to 22 °C? 1 m³ of air has a mass of 1.26 kg. The specific heat of air is 1010 J/kg/K. Express your answer in kilowatt hours.

6. Calculate the amount of heat in joules required to raise the temperature of 100 g of aluminium from 10 °C to 710 °C. The specific heat of aluminium is 915 J/kg/K.

7. Forty litres of water are heated from 8 °C to 78 °C, the efficiency of the operation being 70%. How much heat is required? The specific heat of water is 4187 J/kg/K.

8. The rate of flow of water through a water-cooled motor is 0.2 l/s, and the inlet and outlet temperatures are 10 °C and 20 °C, respectively. At what rate is heat being removed from the motor? The specific heat of water is 4187 J/kg/K.

9. An electric-arc furnace is used to raise the temperature of 4000 kg of iron from 12 °C to 812 °C, the overall efficiency of the furnace being 40%. What energy input in kilowatt hours is required? The specific heat of iron is 497 J/kg/K.

10. An electric water heater is 80% efficient and consumes energy at the rate of 2000 J/s. If the heater initially contains 10 litres of water at 12 °C, what will be the temperature after 10 min of heating? The specific heat of water is 4187 J/kg/K.

11. The input power to a furnace for heating copper is 10 kW, and the furnace is 39.7% efficient. By how much will the temperature of 150 kg of copper have increased after 30 min? The specific heat of copper is 397 J/kg/J.

12. A 2 kW heater is switched on in a room 5 m square and 3 m high, when the air temperature is 70 °C. If 70% of the heat provided is lost, what will be the air temperature after 20 min? 1 m³ of air has a mass of 1.26 kg, and the specific heat of air is 1010 J/kg/K.

13. How long will it take a 1.5 kW heater to raise the temperature of 10 l of water by 60 °C if the heater is 100% efficient? The specific heat of water is 4187 J/kg/K.

14. An instantaneous-type water heater is required to provide a continuous flow of water of 2 l/min at a temperature of 85 °C with an inlet water temperature of 10 °C. What must be the electrical loading in kilowatts if the heater is assumed to be 100% efficient? The specific heat of water is 4187 J/kg/K.

15. A furnace for lead casting contains 120 kg of lead and is heated by a 6 kW element. The initial temperature of the lead is 20 °C. What temperature will the lead reach after 1 h if the furnace is 25% efficient? The specific heat of lead is 129 J/kg/K.

7.9 Multiple-choice exercises

7M1 Heat is
(a) the difference between hot and cold
(b) temperature
(c) a form of energy
(d) something which is responsible for burning

7M2 Temperature is measured using
(a) a thermometer
(b) an ammeter
(c) a sensitive human elbow
(d) a thermistor

7M3 A temperature in kelvin can be converted to one in the Celsius scale by
(a) subtracting 32, multiplying by 5 and dividing by 9
(b) adding 273
(c) using °C after the value instead of K
(d) subtracting 273

7M4 The specific heat capacity of a substance is
(a) how hot it becomes when a given amount of heat is added
(b) the heat energy required to raise the temperature of 1 kg of the material by 1 K
(c) heat energy times mass times temperature change
(d) the heat needed to cool 1 l of water through 10 °C

7M5 A 6 kW instantaneous heater raises the temperature of the water flowing through it from 10 °C to 54 °C. If the specific heat capacity of water is 4187 J/kg/K, the rate of water flow in litres/min is
(a) 1.59
(b) 1.95
(c) 30.7
(d) 0.512

7M6 Most of the energy received by the planet Earth is
(a) by radiated heat from the Sun
(b) a result of nuclear power stations
(c) due to burning coal
(d) a result of tidal action

7M7 The operation of most thermometers is a result of
(a) heat convection
(b) electric heating
(c) heat radiation
(d) thermal expansion

7M8 A soldering iron relies for operation on
    (a) soldering flux
    (b) thermal conduction
    (c) operator skill
    (d) thermal convection

7M9 An immersion heater distributes heat throughout a water storage tank as a result of
    (a) thermal conduction
    (b) lagging
    (c) thermal convection
    (d) thermal radiation

Chapter 8

Magnetism

Magnetism is a form of energy that can cause attraction or repulsion between certain materials. Among many other applications, it can be used to operate switches, drive machines and lift metallic loads. It is also closely bound to electricity. A magnetic field is created when current flows, and, in turn, electricity can be generated using a magnetic field. This section looks at that relationship as well as examining the properties and electro-technological applications of this strange energy.

8.1   Permanent magnets

It is the spin of electrons as they orbit the nucleus of an atom that creates a magnetic field. The internal atomic structure of certain materials enables a magnetic field to remain in place even after an external magnetic force has been removed. These are considered to be permanent magnets, defined as materials that:

- are inherently magnetic
- can be magnetised permanently

In some materials, the presence of an external magnet will cause electrons to align and form domains. Domains are groups of atoms with aligned electrons and these groups retain their magnetic strength without the influence of an external magnetic field. Some metals, such as iron, cobalt and nickel, are considered to be hard ferromagnetic materials, which means that it is difficult to realign their domains to become non-magnetic once they have formed.

Hard ferromagnetic materials are often used as permanent magnets. These are usually weaker than the artificial man-made type. They still have their uses, however. Before the advent of the compass, for example, a magnetic rock known as a lodestone was used to find North and aid navigation for sea voyages. A modern use for permanent magnets is in small generators.

Permanent-magnet materials have been the subject of intense development over the years. Modern versions are resilient to the most adverse conditions of temperature, mechanical ill-treatment and demagnetising fields. The base materials for these magnets are the three main magnetic elements:

- Iron
- Nickel
- Cobalt

Other elements can be added in. For example:

- Tungsten
- Carbon
- Manganese
- Chromium
- Copper
- Aluminium

Modern permanent magnets are often too hard to drill or machine and must be cast in their final correct shapes. Intricate magnetic circuits often use a small permanent magnet at the centre of soft-iron extension pieces. Although modern permanent magnets are unlikely to change their properties during normal use, they can sometimes be damaged magnetically when removed from their surrounding magnetic circuit. If it is necessary to remove a permanent magnet, the following precautions should be observed.

- Complete the magnetic circuit using a mild-steel 'keeper'.
- Do not slide the keeper against the magnet; remove it with a direct pull.
- Do not touch the magnet with steel tools, such as screwdrivers or spanners.
- Do not put the magnet on a steel-topped workbench.
- Keep the magnet in a non-magnetic tray or box, well separated from others.
- Do not allow people to play with the magnet.

8.2   Electromagnets

Electromagnets are created by surrounding a **ferrous** material with an electrical field. This has the effect of aligning all the molecules in an identical north-south orientation. This is usually achieved practically by winding an insulated conductor around a ferrous former. The former is called the iron core.

> **KEY TERM: Ferrous Metals – Ferrous comes from the symbol F, given to the element iron. Ferrous metal can be pure iron or an alloy that has iron as one of its components.**

Unlike permanent magnets, electromagnets only retain their magnetic fields while current is flowing through the surrounding conductors. This effect is utilised for a number of applications such as electrical control circuits.

There are at least four main types of wound electromagnets:

- Coil – This provides the magnetic force for the electromagnetic switches in solenoids, circuit breakers and contactors, as well as non-electrical applications such as valves in gas and fluid systems.
- Winding – This provides a magnetic field that is used to power electric motors and transformers, as well as an integral part of electrical production in generators.

- Electromagnet – It is used for many applications such as cranes employed to lift scrap metal.
- Heating – The heat produced by electromagnetic energy is used in domestic induction hobs and industrial furnaces.

8.3 Magnetic fields

Magnetic energy manifests as a field around the magnetic material. This magnetic field extends outwards into the space surrounding the magnet. It grows weaker as the distance from the magnet increases. There are several methods of detecting the presence of a magnetic field, although it is invisible and does not affect human senses. Theoretically, the magnetic field extends for considerable distances, but in practice it will combine with the fields of other magnets to form a composite, so that the effects of most magnets can only be detected quite close to them.

Flux lines

Because we cannot see, feel, smell or hear it, a magnetic field is difficult to represent. A 'picture' of a magnetic field is often useful in deciding what its effects will be, and to do this we consider that a magnetic field is made up of imaginary lines of flux that have the following properties.

- They always form complete, closed loops.
- They never cross one another.
- They have a definite direction.
- They try to contract as if they were stretched elastic threads.
- They repel one another when lying side by side and having the same direction.

It is important to remember lines of magnetic flux do not exist as they are usually drawn. Representing them in this way gives us a method of understanding the behaviour of a magnetic field.

Magnetic poles

We know that the Chinese experimented with crude permanent magnets hundreds of years ago and discovered that a freely suspended magnet will come to rest with a particular part (one end, if it is a bar magnet) pointing north. This is the north-seeking pole or north pole of the magnet, and the other end is the south-seeking pole or south pole. If two magnets, with their poles marked, are brought together, the basic laws of magnetic attraction and repulsion can easily be demonstrated. These are as follows:

- Like poles repel. For instance, two north poles or two south poles will try to push apart from each other.
- Unlike poles attract. For instance, a north pole and a south pole will attract one another.

The Earth behaves as if there is a huge bar magnet inside it.

- The south pole of the magnet is near the geographic North Pole.
- The north pole at the geographic South Pole.

The positions of the magnetic north and south poles of the Earth vary slowly with time, for reasons not yet fully explained. As far as we can tell, however, the Earth is a truly permanent magnet.

Attraction and repulsion

The pattern of the lines of magnetic flux around a permanent bar magnet is represented in Figure 8.1. This shows that the direction of the lines of magnetic flux (property 3) is from the north pole to the south pole outside the magnet. These lines of magnetic flux obey the five rules given earlier. Their curved shapes outside the magnet are a compromise between properties 4 and 5.

Figure 8.2 shows the flux pattern due to two magnets with unlike poles close together. Lines of magnetic flux try to contract (property 4), and the magnets try to pull together.

Figure 8.3 shows the magnetic flux pattern due to two magnets with like poles close together. Since magnetic flux lines running side by side with the same direction repel, the two poles try to push apart.

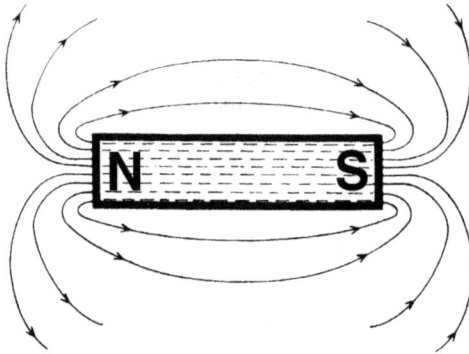

Figure 8.1   Magnetic field due to bar magnet

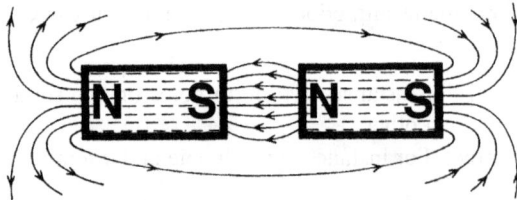

Figure 8.2   Magnetic field due to two bar magnets with unlike poles adjacent

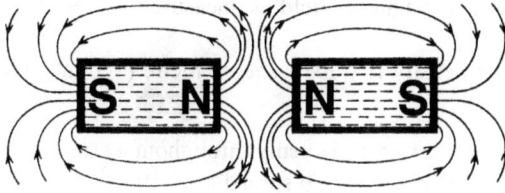

Figure 8.3   Magnetic field due to two bar magnets with like poles adjacent

These explanations indicate the neat way in which the results of magnetic fields can be forecast by the use of lines of magnetic flux. These lines can be plotted for an actual magnetic field in two ways:

1. Cover the magnet system concerned with a sheet of paper and then sprinkle it with iron filings, which will set themselves in the magnetic flux pattern.
2. Take a plotting compass, which consists of a miniature magnetic needle pivoted within a clear container.
   (a) The compass is placed on the paper, and the position of its north and south poles is marked with a pencil.
   (b) It is then moved and its position adjusted until its south pole is at the mark indicating the previous position of the north pole.
   (c) The new position of the north pole is marked and the process repeated as often as is necessary.
   (d) The line of marks is joined to form a line of magnetic flux. As many lines as are required can be produced in a similar way.

8.4   Units of magnetic flux

Magnetic flux (symbol, ϕ; SI unit, weber Wb)

Lines of magnetic flux can be counted and used to measure the quantity of magnetism.

- The general symbol for magnetic flux is ϕ (Greek letter 'phi').

- The unit of magnetic flux is called the weber (pronounced 'vayber' and abbreviated to Wb).

Flux density (symbol, B; SI unit, tesla T)

The weber is a measure of a total amount, or quantity, of magnetic flux, and not a measure of its concentration or density. Flux density is very important in some machines and depends on the amount of magnetic flux concentrated in a given cross-sectional area of the flux path.

- The strength of a magnetic field is measured in terms of its flux density (symbol B).
- The unit of flux density is webers/square metre (Wb/m²), more usually called tesla (T).

1 Wb of magnetic flux spread evenly throughout a cross-sectional area of 1 m² results in a flux density of 1 T. 1 Wb spread over 10 m² will give a flux density of 0.1 T.

$$B = \frac{\phi}{A}$$

where B = magnetic flux density, T; ϕ = magnetic flux, Wb; and A = cross-sectional area of flux path, m². Flux areas of the order of square metres are only met in the largest machines, but small quantities of flux in small areas can give rise to high magnetic flux densities.

Example 8.1
The magnetic flux per pole in a DC machine is 2 mWb, and the effective poleface dimensions are 0.1 m × 0.2 m. Find the average flux density at the poleface.

Convert m Wb to Wb. $\dfrac{2}{1000} = 0.002$ Wb

Area $= 0.1 \times 0.2 = 0.02$ m²

$$B = \frac{\phi}{A} = \frac{0.002}{0.02} = 0.1 \text{ T}$$

Example 8.2
A lifting electromagnet has a working magnetic flux density of 1 T, and the effective area of one poleface is a circle of diameter 70 mm. What is the total magnetic flux produced?

Effective poleface area $= \dfrac{\pi d^2}{4} = \dfrac{\pi \times 70^2}{4} = 3850$ mm²

Convert to m² $\dfrac{3850}{\left(10 \times 10^6\right)} = 0.00385$ m²

Transpose $B = \dfrac{\phi}{A}$ to make ϕ the subject

$\phi = BA = 1 \times 0.00385$ webers $= 0.00385$ webers $= 3.85$ mWb

---

8.5 Electromagnet properties

Conductor magnetic fields

When electrical current flows, a magnetic field forms about the conductor. The flux is shown as a corkscrew, which is clockwise-orientated in relation to the direction of current flow.

Figure 8.4 shows a plot of the magnetic flux around a current-carrying conductor. Figure 8.5 shows these magnetic fields as viewed along the cross-section of a conductor. The symbols that indicate direction of current flow are:

- A cross – current flowing away from the viewer.
- A dot – current flowing towards the viewer.

These symbols are based on a dart or arrow. If you throw a dart you will see the x-shape of the flights. If a dart is flying towards you, then you will see the tip of its sharp point as a dot.

Another method for remembering the direction of magnetic flux about a conductor is by using the screw rule. This states that if a normal right-hand thread screw is driven along the conductor in the direction taken by the current, its direction of rotation will be the direction of the magnetic field.

It is quite common in electrical work for several current-carrying conductors to be installed side by side in a cable or conduit. Figure 8.6 shows that the magnetic field in the space between the conductors may be intense. BS7671 states that when

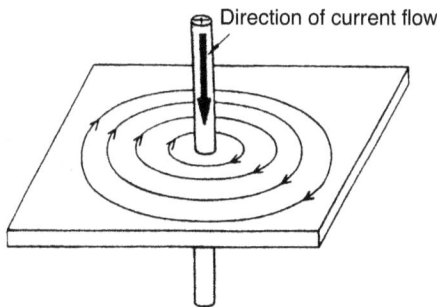

Figure 8.4   Plotting magnetic field due to current-carrying conductor

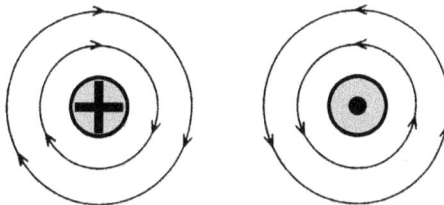

Figure 8.5   Direction of magnetic field around current-carrying conductor

cables are carried in metal conduit or trunking, there should be a mix of opposing currents. This reduces heat and the chance that the combined magnetic fields of the cables could cause vibration and loosening of joints in the **containment.** The mix does not have to be 50/50. This means, for example, that you should *never* install all the line conductors in one conduit and all the neutrals in the other (Figure 8.6).

> **KEY TERM: Containment – Wiring systems that use conduit or trunking as a route for cables. They also provide protection to the cables**

The solenoid

The strength of the magnetic field around a conductor depends on the current. Even at high currents, it is comparatively weak, however. Increasing the number of conductors will increase the strength of the field. This can be achieved using a single conductor formed into a tight spiral. This is called a coil and is the basis of the solenoid.

Figure 8.7(b) is a cross-section of the solenoid showing how the individual magnetic fields merge to form a stronger field. This field is similar to that of the permanent bar magnet (Figure 8.1). The strength of the magnetic field produced depends on:

- the current
- the number of turns used. Additional turns may be wound on top of the first layer to form a multilayer coil.

The north-south rule can be seen in Figure 8.8. This proved by the Right Hand Grip Rule (Figure 8.9). If you could grip the coil with your hand, sliding your

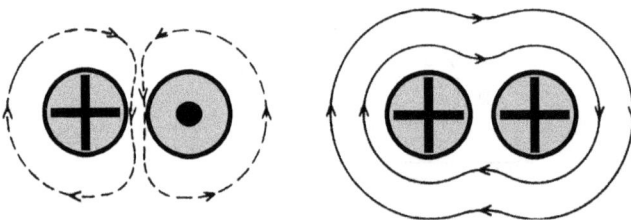

Figure 8.6   Magnetic field due to two current-carrying conductor

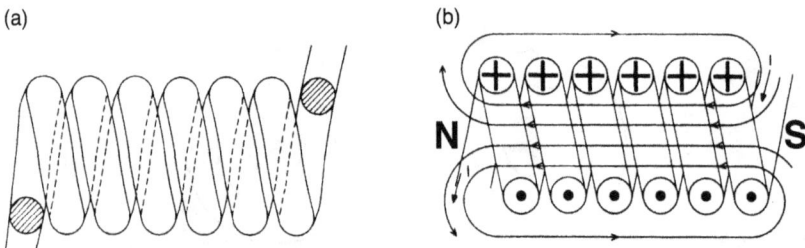

Figure 8.7   Solenoid and its magnetic field

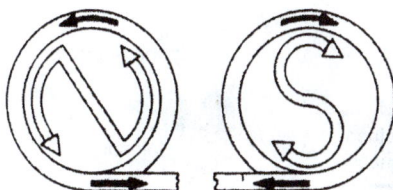

Figure 8.8   NS rule

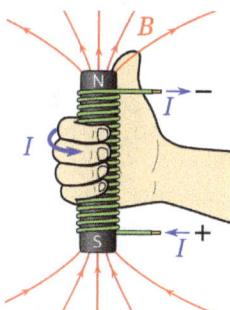

Figure 8.9   Right Hand Grip Rule. Image credit: MikeRun, via Wikimedia Commons, Public domain.

fingers around the turns in the same direction as the current flow, then your thumb will indicate the north pole of the magnetic field.

The solenoid is the electrical equivalent of the permanent magnet, but is often more versatile. For instance, the flux and flux density can be altered by increasing or decreasing the solenoid current or reduced to zero by switching the current off. Since the solenoid is simply a coil of conductor, its size and shape can be constructed to almost any requirements. The electromagnet, based on the solenoid, is at the heart of many items of electrical equipment, for example:

- Relays
- Contactors
- Electric motors
- Generator
- Transformers

8.6   Calculations for air-cored solenoids (*symbols magnetomotive force, MMF and H; SI units, ampere-turns or ampere/metre At/m or Am*)

An air-cored solenoid consists of a conductor coiled around a hollow, non-magnetic former (Figure 8.10). This can be plastic or some other insulating material, or it can be air. In this case, the magnetic field central to the solenoid's operation is created in the air and not in the former.

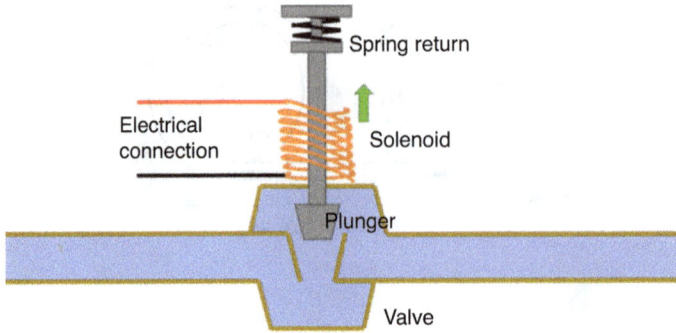

*Figure 8.10   A solenoid valve. Image credit: Joey Corbett, via Wikimedia
Commons, CC BY-SA 4.0.*

If we multiply the number of turns of a solenoid by the current it carries, we arrive at the magnetomotive force (MMF) of the solenoid, This is measured in ampere-turns (At) or, more strictly, in amperes (A):

magnetomotive force = amperes × turns MMF = IN

For instance, two of the methods of producing an MMF of, say, 1000 At are:

- 1 A flowing in a 1000-turn coil
- 10 A flowing in a 100-turn coil.

The magnetomotive force obviously affects the flux set up by the solenoid, but so too does the length of path taken by individual lines of magnetic flux. For instance, a 1000-turn coil wound in four layers and having a length of 0.1 m will set up more flux for a given current than a 1000-turn coil wound in one layer over a length of 0.4 m. The path taken by the lines of magnetic flux will be longer in the 0.4 m version, so that the magnetomotive force is 'stretched' over a greater distance.

The magnetomotive force applied to each metre of the path taken by the lines of magnetic flux is called the magnetising force (symbol H) and is measured in ampere turns per metre (At/m) or, more strictly, in amperes per metre (A/m)

$$H = \frac{\text{amperes} \times \text{turns}}{\text{length of flux path in metres}} \quad \text{or} \quad H = \frac{IN}{l}$$

where H = magnetising force, At/m or A/m; I = solenoid current, A; N = number of turns of conductor on solenoid; l = mean length of magnetic flux, m. The path taken by the lines of magnetic flux is often referred to as the magnetic circuit.

Example 8.3
A solenoid wound with wire having an overall diameter of 1 mm, must have 12 layers of winding, each 100 mm long. Calculate how much current must flow in the coil to give a magnetising force of 3000 At/m in a magnetic coil circuit of

average length 0.25 m.

$$\text{Number of turns per layer} = \frac{\text{Solenoid length}}{\text{Wire diameter}} = \frac{100}{1} = 100 \text{ turns}$$

Total turns $= 12 \times 100 = 1200$ turns

$$H = \frac{IN}{l}; \quad \text{therefore, } I = \frac{Hl}{N} = \frac{3000 \times 0.25}{1200} = 0.625 \text{ A}$$

Permeability (symbol μ_0; unit, henrys/meter H/m)

Permeability is a measurement of the level of magnetism induced into a material when it is exposed to a magnetic field.

Note: Henry is a unit of inductance, i.e. the ability of a magnetic field to affect another material or to produce electricity.

Permeability of free space

When a current flows in a coil, the resulting magnetising force sets up a magnetic flux. The amount of flux set up in air and other materials that are not attracted by a magnet will depend directly on the applied magnetising force. For instance, the extremely high magnetising force of one million ampere turns per metre would produce a flux density of $4\pi/10$ or 1.257 T within the solenoid. This is illustrated by the ratio:

$$\frac{B}{H} = \frac{4\pi/10}{10^6} \quad \text{or } 4\pi \times 10^7$$

This constant relationship between flux density and magnetising force for air is called the permeability of free space, and is given the symbol μ_0 (μ is the Greek letter 'mu').

$$\mu 0 = \frac{4\pi}{10^7}$$

Example 8.4

An air-cored solenoid is in the form of a closed ring (or toroid) of mean length 0.2 m and cross-sectional area 1000 mm². It is wound with 1000 turns and carries a current of 2 A. Find the:

(a) Magnetic-flux density
(b) Total magnetic flux produced within the solenoid

(a) *Magnetic Flux Density*

$$H = \frac{IN}{l} = \frac{2 \times 1000}{0.2} = 10,000 \text{ At/m}$$

$$\frac{B}{H} = \mu_0; \quad \text{therefore, } B = \mu_0 H = \frac{4\pi}{10^7} \times 10,000 = 0.01257 \text{ T} \quad \text{or} \quad 12.57 \text{ mT}$$

(b) *Total Magnetic Flux*

$$\Phi = BA = \frac{0.01257 \times 1000}{10^6} \text{ (note: } 1 \text{ m}^2 = 10^6 \text{ mm}^2)$$

$$= 0.00001257 \text{ Wb} \quad \text{or} \quad 12.57 \text{ } \mu\text{Wb}$$

Notice the low magnetic-flux density and total flux set up by the comparatively high magnetising force.

Example 8.5
A solenoid of 10,000 turns is wound on a brass ring of mean diameter 100 mm and cross-sectional area 0.01 m². How much current must flow to the solenoid to produce a total flux of 5 mWb?

Brass is a non-magnetic material, so the rules given will apply.

$$B = \frac{\Phi}{A} = \frac{0.005}{0.01} = 0.5 \text{ T}$$

$$\frac{B}{H} = \mu_0; \text{ therefore, } H = \frac{B}{\mu_0} = \frac{IN}{l}$$

$$H = \frac{0.5 \times 10^7}{4\pi}$$

Also

$$H = \frac{I \times 10,000}{\pi/104\pi} \quad I = \frac{0.5 \times 107 \times \pi}{4\pi \times 10,000 \times 10} = 12.5 \text{ A}$$

8.7   Effect of iron on magnetic circuits

Relative and absolute permeability

If a solenoid has an iron core, the magnetic flux produced will be higher than if it has a hollow, air-core, even if the current supplied to the solenoid is the same for both.

The ratio of flux produced by a solenoid with a magnetic core to flux produced by the same solenoid with an air-core (the current being the same in both cases) is called the relative permeability (symbol μ_r) of the material under these conditions.

- The value of the relative permeability of a nonmagnetic material is unity (one).
- The relative permeability for magnetic materials vary between wide limits. Typical figures for common magnetic materials range from 150 (approx.) to 1200

(approx.). This value is not a constant for a given material, since it depends on the magnetising force applied. Taking relative permeability into account.

$$\frac{B}{H} = \mu_0\mu_r$$

The product $\mu_0\mu_r$ is called the absolute permeability of the material under the given conditions.

In many applications, a magnetic circuit has a small airgap in it. In some, such as contactors or machines, this is because free movement must be possible between different parts of the magnetic circuit. In other machines, poorly fitting joints in the magnetic circuit have the same effect as an airgap. Even a small airgap may have a disproportionate effect on a magnetic circuit. Example 8.6 illustrates the advantage of using a magnetic material for a magnetic circuit, and the reduction of flux that occurs if an airgap is introduced

---

Example 8.6

A solenoid is made to be identical with that described in Example 8.4, but is wound on a wrought-iron core. When the solenoid current is 0.2 A, relative permeability of the wrought iron is 500. Calculate:

(a) The flux density in the ring.
(b) Total flux in the ring.
(c) What would be the effect on flux and flux density if a radial sawcut were to be made in the wrought-iron core without disturbing the magnetising coil?

(a) *Flux Density in the Ring*

$$H = \frac{IN}{l} = \frac{0.2 \times 1000}{0.2} = 1000 \text{ At/m}$$

$$\frac{B}{H} = \mu_0\mu_r; \text{ therefore, } B = \mu_0\mu_r H$$

$$B = \frac{4\pi}{10^7} \times 500 \times 1000 = 0.628 \text{ T}$$

(b) *Total Flux*

$$F = BA = 0.628 \times 0.001 = 0.000628 = 0.628 \text{ mWb}$$

(c) *Effect of a Gap in Ring*

The introduction of an airgap into the wrought iron magnetic circuit will reduce the magnetic flux and flux density set up by the magnetising coil. Calculations of flux values in a case of this sort are complicated, but it can be shown that if the sawcut is only 3 mm wide, flux density will be reduced to 0.0714 T and flux to 0.0714 mWb.

The results of this example are worthy of close examination, since they show some of the principles of magnetic-circuit design. Part (a) shows that the substitution of a magnetic material as the core of the solenoid increases the flux and flux density by 50 times, even though the magnetising current is one-tenth its previous value.

The answer to part (b) shows that the airgap (length 3 mm) makes the establishment of magnetic flux nearly ten times as difficult as when the iron path was complete. This is despite the fact that 98.5% of the circuit (197 mm) still consists of iron.

8.8   Practical applications of electromagnetism

Relays

As mentioned previously, a relay is a switch operated by the energising and dee-nergising of an electromagnetic coil (Figure 8.6). This enables one circuit to operate another. It also allows a very low current to operate a high-power circuit, because a relay coil only requires a small current.

Low current automatic controls such as thermostats and timers can also be wired into the coil circuit. Some relays have a large number of contacts, and can be used in complicated circuits for a wide variety of switching purposes. As in other applications, however, electronic switching is becoming common. Newer types of relay are now solid state, using thyristors or triacs rather than electromagnetic coils.

A relay usually has an operating coil wound on a magnetic circuit with a moving section, or armature, held open by a spring (Figure 8.11). When the

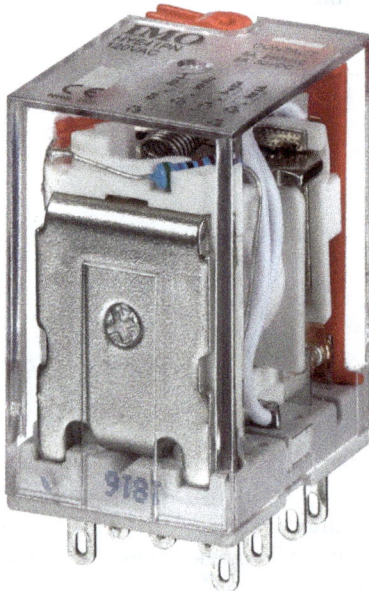

Figure 8.11   A typical relay (by kind permission of IMO Technology)

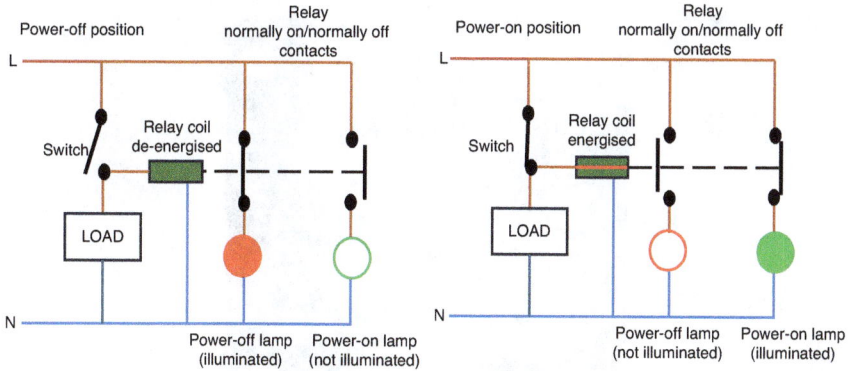

Figure 8.12 Simple example of how a normally open/normally closed (NO/NC) relay contact can be used to operate power-on and power-off lamps. Current is directed to the off lamp by the NC contact when the circuit is de-energised, then to the on lamp via the NO contact when the circuit is de-energised.

operating current is switched on and sets up a magnetic field, the armature is closed and operates the contacts.

Control circuit switches can be one of the following. Note that the description, *normally*, applies when no current flows:

- Normally open – a conventional on/off switch that closes when the relay is energised.
- Normally closed – allows current flow until the relay is energised and the switch opens.
- Normally open/normally closed – a changeover switch, like the points on a railway track. It would allow current to flow through one route when the relay is de-energised then open that circuit and close another when the relay is energised. It could be used to operate power on-and-off lights (Figure 8.12).

Contactors

Contactors are simply very large relays that enable a heavy load to be switched on and off with a small operating current (Figure 8.13). A few applications are as follows.

- Motor starting: The contactor coil, and hence the motor, is controlled by 'stop' and 'start' pushbuttons. The contactor also ensures that the supply to the motor is disconnected in the event of a supply failure, so that the motor cannot restart automatically when the supply is restored (known as no volt [or low volt] protection).
- Timeswitch contactor: A heavy heating load can be switched by a timeswitch with low-rated contacts; the timeswitch controls the coil of the contactor, which controls the load.

Figure 8.13   Four-pole contactor (by kind permission of IMO Technology)

- Lighting loads too heavy to be operated by a conventional light switch. The low current light switch operates the contactor coil, which, in turn opens and closes the higher current contacts through which the actual lighting circuit is wired.

Loudspeakers

The loudspeaker converts electrical signals into sound waves.

- A typical loudspeaker has a magnetic circuit comprising a permanent magnet, and completed by soft-iron pole pieces. This results in a strong magnetic flux in a short cylindrical airgap.
- A moving coil, called the speech or voice coil, is suspended from the end of a paper or plastic cone so that it lies in the gap (Figure 8.14).
- Alternating currents in the coil push it up and down, vibrating the cone and producing sound waves.

The quality of a loudspeaker depends on:

- The strength of the magnetic field in the gap
- The lightness of the coil, which is often wound of aluminium wire
- The size of the cone. A large cone produces low-frequency notes well, and a small cone deals better with high frequency notes. Some loudspeakers have two cones to enable them to reproduce notes faithfully over a wide range of frequencies

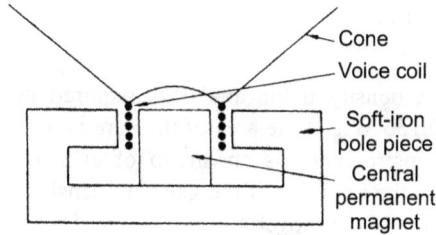

Figure 8.14 A typical loudspeaker

8.9 Summary of formulae for Chapter 8

Flux Density

$$B = \frac{\Phi}{A} \quad \Phi = BA \quad A = \frac{\Phi}{B}$$

where B = magnetic-flux density T (Wb/m^2); Φ = magnetic flux, Wb; and A = cross-sectional area of flux path, m^2.

Magnetomotive Force

$$H = \frac{IN}{l} \quad I = \frac{Hl}{N} \quad N = \frac{Hl}{I}$$

where H = magnetising force (ampere turns per metre, or amperes per metre); I = magnetising current, A; N = number of magnetising coil turns; l = mean length of magnetic-flux path, m.

Permeability of Free Space

$$\mu_0 = \frac{4\pi}{10^7} \quad \mu_0 = \frac{B}{H} \quad B = \mu_0 H \quad H = \frac{B}{\mu_0}$$

where μ_r is the relative permeability of the magnetic material under the given conditions.

8.10 Exercises

1. It is required to determine the magnetic field surrounding a bar magnet. Give a brief description of a laboratory procedure to do this.
2. Draw a diagram to show a solenoid wound over an iron core. Mark on the diagram the magnetic lines of flux resulting from current in the solenoid. Show a direction for the current, and give the resulting direction of the flux lines.
3. Draw circles to represent the cross-sections of two conductors lying side by side. Mark the directions of currents in the conductors, and sketch the resulting magnetic field if the two conductors carry currents

(a) in the same direction and

(b) in opposite directions

4. What is the flux density if the total flux required in the core of a power transformer is 0.156 Wb. If the area of the core is 0.12 m²?

5. A moving-coil instrument has an airgap of effective cross-sectional area 56 mm². What is the total flux if the gap flux density is 0.12 T?

6. A coil of 400 turns wounds over a wooden ring of mean circumference 0.5 m and uniform cross-sectional area 800 mm² carries a current of 6 A. Calculate:

(a) Magnetising force

(b) Total flux

(c) Flux density

7. If an airgap of length 20 mm is introduced into a magnetic circuit of cross sectional area 0.009 m² that is carrying a flux of 1.2 mWb, what extra magnetomotive force will be required to maintain the flux?

8. Calculate the flux and flux density set up by the solenoid of Exercise 6 if the wooden ring is replaced by one of magnetic material having identical dimensions and a relative permeability of 200 at this flux density.

9. A coil of insulated wire of 500 turns and of resistance 4 Ω is closely wound on an iron ring. The ring has a mean diameter of 0.25 m and a uniform cross sectional area of 700 mm². Calculate the total flux in the ring when a DC supply at 6 V is applied to the ends of the winding. Assume a relative permeability of 550. Explain the general effect of making a small airgap by cutting the iron ring radially at one point.

10. Make sketches indicating current direction and resulting magnetic field when current flows in

(a) a straight conductor

(b) a solenoid about 40 mm in diameter and 50 mm long

(c) a similar solenoid with a 20-mm diameter steel core 50 mm long

(d) a solenoid with a steel core in the shape of a closed ring.

11. Figure 8.15 shows a coil of wire wound onto an iron core with an airgap in one limb. How would the magnetic flux in the gap be affected by an increase in

(a) the current in the coil

(b) the number of turns of wire

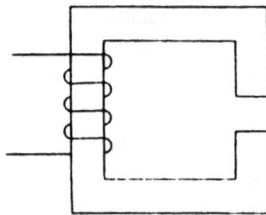

Figure 8.15   Diagram for Exercise 11

(c) the length of the iron path
(d) the length of the gap
(e) the permeability of the iron

8.11   Multiple-choice exercises

8M1 A space in which magnetic effects can be detected is called
   (a) permeability
   (b) a magnetic field
   (c) magnetic flux
   (d) lines of force

8M2 Lines of magnetic flux are considered to have a number of properties. Identify the property in the following list which is INCORRECT
   (a) they have a definite direction
   (b) they always form complete closed loops
   (c) they never cross one another
   (d) they attract each other when lying side by side and having the same direction.

8M3 If the south-seeking poles of two magnets are brought close together, they will
   (a) repel each other
   (b) be unaffected
   (c) pull together
   (d) attract each other

8M4 Lines of magnetic flux are considered to have the direction
   (a) towards a north pole
   (b) north to south outside a magnet
   (c) towards a south pole
   (d) south to north outside a magnet

8M5 Magnetised soft iron may lose its magnetism if
   (a) immersed in water
   (b) placed in a drawer
   (c) heated or hammered
   (d) bent

8M6 The shape of a magnetic field can be found by using
   (a) a plotting compass or iron filings
   (b) an Ordnance Survey map
   (c) a land surveyor
   (d) an ammeter

8M7 The symbol used to represent magnetic flux density is
   (a) T
   (b) D

(c) Wb

(d) B

8M8 If a flux density of 0.16 T appears at the 20 cm × 10 cm poleface of a machine, the total flux is

(a) 32 Wb

(b) 3.2 mWb

(c) 8 T

(d) 312.5 mWb

8M9 A current flowing towards the viewer in a cross-sectioned conductor is shown by

(a) clockwise concentric circles

(b) a cross

(c) a dot

(d) anticlockwise concentric circles

8M10 If two adjacent conductors are carrying current in the same direction, they will both experience a force that

(a) pushes them apart

(b) tries to move them lengthways

(c) heats them up

(d) pulls them together

8M11 A solenoid is

(a) a kind of heat lamp for browning the skin

(b) a type of permanent magnet

(c) a coil of wire that sets up a magnetic field when is carries an electric current

(d) part of a boot

8M12 The formula for magnetising force is

(a) $H = \frac{IN}{l}$

(b) $B = \frac{\mu_0 \mu_r}{H}$

(c) $B = \frac{H}{A}$

(d) $\mu_0 = 4\pi \times 10^{-7}$

8M13 A magnetic circuit has a mean length of 0.4 m, a cross-sectional area of 120 mm^2 and is made of copper. If wound with 2500 turns of wire carrying a current of 120 mA, the total flux set up in the copper will be

(a) 750 A/m

(b) 0.113 μWb

(c) 7.85 Wb

(d) 0.036 Wb

8M14 The formula relating absolute permeability, relative permeability, magnetic flux density and magnetising force is

(a) $H = \frac{IN}{I}$

(b) $\mu_0\mu_r = BH$

(c) $A = \frac{H}{B}$

(d) $B = \frac{\mu_0\mu_r}{H}$

Chapter 9

Electromagnetic induction

9.1 Introduction

Scientists have known of the existence of electric current and the magnetic field for many centuries, but the connection between the two was not understood. Thanks to the work of Oersted, Faraday and others, electricity and magnetism are now considered to be inseparable.

In Chapter 8, it was shown that the flow of current in a conductor gives rise to a magnetic field. In this chapter, it will be shown that under certain conditions a magnetic field can be responsible for the flow of an electric current. This effect is known as electromagnetic induction.

9.2 Dynamic induction

The word 'dynamic' suggests force and movement. In dynamic induction, a conductor is moved *across* the lines of flux in a magnetic field. It is the change in flux strength that causes induction to take place. This means that if you move the conductor *along* the lines of flux, there will be no change in flux strength and, therefore, no induction.

Induction can be effected in three ways:

- Move the conductor across the lines of flux.
- Move lines of flux over a stationary conductor.
- Both magnetic field and conductor are stationary, but the flux strength is changed.

If a length of flexible conductor has its ends connected to a sensitive indicating instrument, the needle of the instrument will give a sharp kick when the conductor is moved across a magnetic field (Figure 9.1).

- This simple experiment shows that if a conductor moves across a magnetic field, an EMF is induced.
- This EMF will cause a current to flow if a closed electric circuit exists.
- If a number of conductors are assembled so that they can rotate between the poles of a magnet, we have a simple generator, the principles of which will be explained at a later stage.

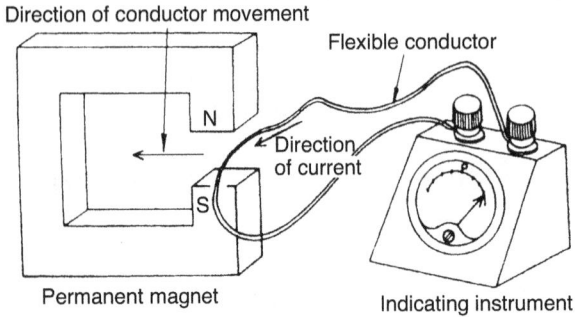

Figure 9.1   Induced EMF in conductor due to its movement in magnetic field

The amount of EMF generated using a magnetic field is affected by three things:

- Conductor length: If the flexible conductor is wound in a loop so that the adjacent sides of two turns pass at the same speed through the same magnetic field, the deflection of the instrument will be twice as great as when the single conductor is used. From this, it follows that induced EMF depends on the length of the conductor subject to the magnetic field.
- Magnetic flux: If the magnet is changed for a stronger one (with greater magnetic flux density between its pole faces), the deflection will again be greater. EMF, therefore, depends on the density of the magnetic field through which the conductor passes.
- Velocity: The deflection of the needle can be shown to depend on the speed, or velocity, of the conductor through the magnetic field. The faster the conductor moves, the greater the deflection.

These three effects can be calculated using the formula:

$$e = Blv$$

where e = induced EMF, V; B = flux density of magnetic field, T; l = length of conductor in the field, m; and v = velocity (speed) of the conductor, m/s.

Note that e will be a steady value only as long as the conductor velocity is constant, and as long as the flux density of the magnetic field remains the same. For this reason, the symbol e, for an instantaneous value, is used. The conductor must move directly across the magnetic field so that its path is at right angles to the magnetic flux.

Example 9.1
A conductor, 300 mm long, moves at a fixed speed of 2 m/s through a uniform magnetic field of flux density of 1 T. What current will flow in the conductor

(a) if its ends are open-circuited?
(b) it its ends are connected to a 12 Ω resistor?

In both cases

$e = Blv = 1 \times 0.3 \times 2 = 0.6$ V

(a) If the ends of the conductor are open-circuited, no current will flow.

(b) $I = \frac{E}{R} = \frac{0.6}{12} = 0.05$A or 50 mA (neglecting conductor resistance).

Example 9.2
A conductor of effective length 0.5 m is connected in series with a milliammeter of resistance 1 Ω, which reads 15 mA when the conductor moves at a steady speed of 40 mm/s in a magnetic field. What is the average flux density of the field?

$E = IR = 0.015 \times 1 = 0.015$ V

$e = Blv$ so $B = \dfrac{e}{lv}$ $B = \dfrac{0.015}{0.5 \times 0.04} = 0.75$ T

An alternative method of calculating induced EMF is based on the definition of the unit of magnetic flux, the weber. If a conductor moving at constant speed cuts a total of one weber of flux in one second, the EMF induced in it will be one volt. Thus, the induced EMF is equal to the average rate of cutting magnetic flux in webers per second; that is

$$\text{volts} = \text{webers/second or } e = \frac{\Phi}{t}$$

Example 9.3
A conductor is moved across a magnetic field, having a total flux of 0.2 Wb, in 0.5 s. What will be the average EMF induced?

$e = \dfrac{\Phi}{t} = \dfrac{0.2}{0.5} = 0.4$ V

9.3   Relative directions of EMF, movement and flux

As well as the strength of voltage and current, polarity can also be changed during electromagnetic induction.

- If the direction of movement of the conductor in the magnetic field (Figures 9.1 and 9.2) is reversed, the EMF induced will have reversed polarity.
- If the magnetic field direction is reversed again, this will again reverse the polarity.

It is often necessary to be able to forecast the direction of one of the three variables (magnetic field, induced EMF and conductor movement) if the other two are

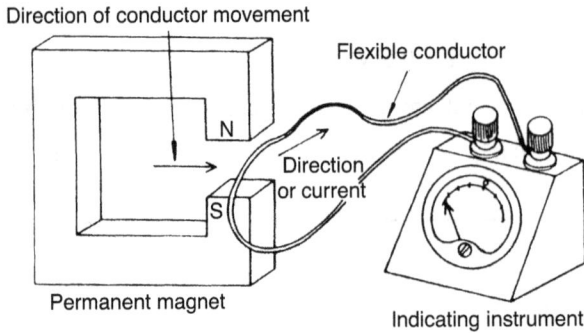

Figure 9.2   Dependence of induced EMF direction on direction of conductor movement relative to magnetic field

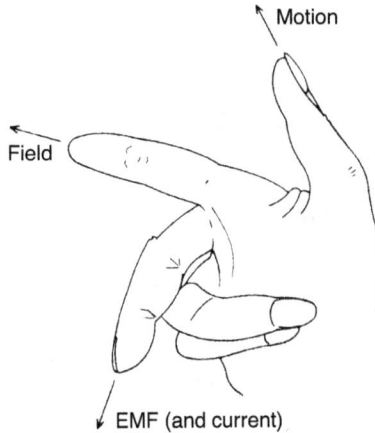

Figure 9.3   Position of right hand for application of Fleming's right-hand rule

known. The easiest of several methods is the application of Fleming's right-hand (generator) rule.

The thumb and first two fingers of the right hand are held at right angles to each other (Figure 9.3). Then, if the:

- First-finger direction is that of the magnetic Field (north pole to south pole across the magnet);
- seCond-finger is that of current as a result of induced EMF, then the thuMb shows the direction of conductor movement in the magnetic field.

Always remember to use the **right hand** for generators.

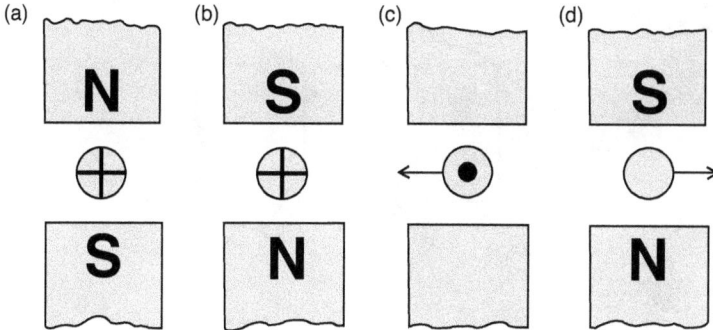

Figure 9.4   Diagram for Example 9.4

Example 9.4
Examine the diagrams of a conductor between a pair of magnetic poles (Figure 9.4), and state

(i) the direction in which the conductor moves for Figure 9.4(a) and (b)
(ii) the polarity of the magnet system for Figure 9.4(c)
(iii) the direction of the induced EMF for Figure 9.4(d)

The answers are

(a) left to right
(b) right to left
(c) north pole at the top
(d) upwards (out of the paper).

9.4   Simple rotating generator

We have already seen that an EMF is induced into a conductor when it is moved across a magnetic field. When the conductor moves along the field, no flux is cut, and no EMF induced.

- If the conductor moves at right angles to the field, flux is cut at the maximum rate and maximum EMF induced.
- If the conductor cuts the field obliquely, the EMF induced lies between the zero and maximum values and depends on the angle at which the flux is cut (Figures 9.5 and 9.6).

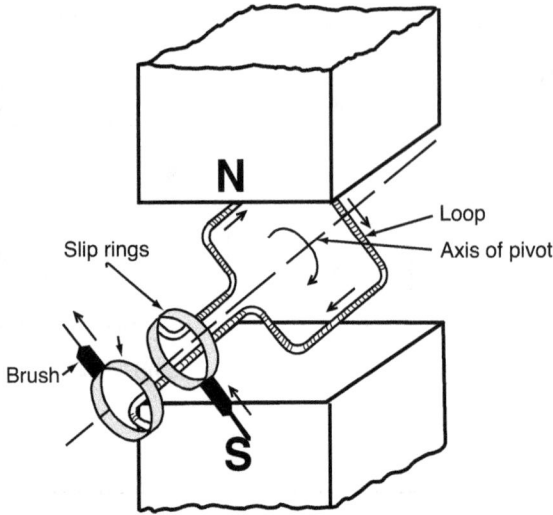

Figure 9.5   Loop connected to slip rings and able to rotate in magnetic field

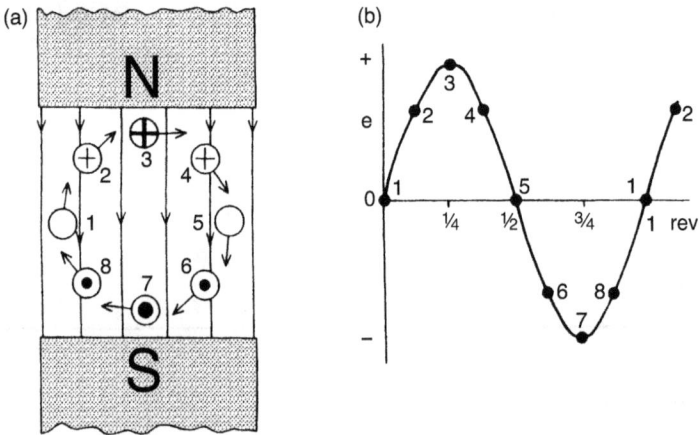

Figure 9.6   Loop rotating between poles of a magnet: (a) section through loop of Figure 9.5, which is shown in several positions and (b) EMF in loop drawn as graph

9.5   Direct-current generator

Direct current is a form of electrical supply in which positive and negative remain constant. In other words, the positive side of the generator's output is always positive and the negative is always negative. A battery provides a DC supply, as do many renewable energy sources.

(a)

(b)

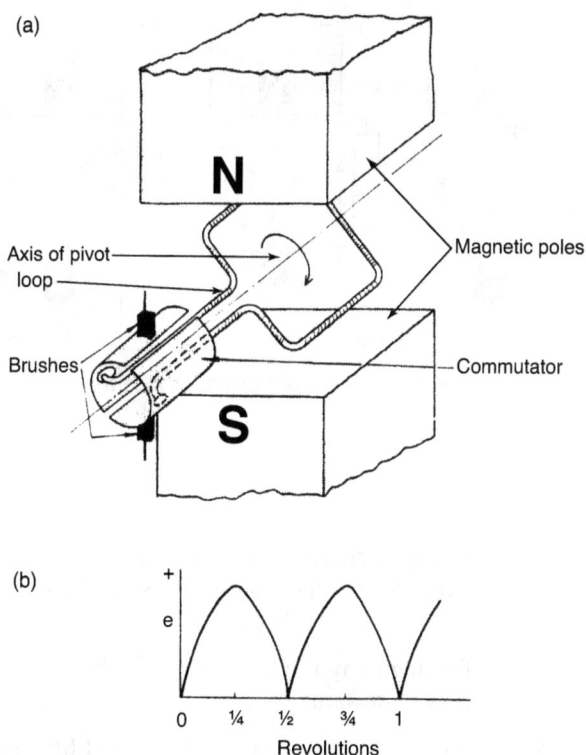

Figure 9.7   *Simple DC generator: (a) loop generator connected to simple commutator and able to rotate in magnetic field and (b) output EMF from system*

The main supply from power stations and out through the grid is alternating current (AC). Many items of electrical equipment need AC to operate. Unlike DC, the polarity of an AC supply constantly changes (Figure 9.6). In the UK, the rate of those changes, or frequency, is 50 times/second (50 hertz). However, there has always been a demand for DC; cranes and railways, for example, and more recently, the countless digital devices now in common use. AC can be converted to DC by means of a rectifier, but it can also be generated at the source.

The key to DC generation is the commutator, which is a segmented metal drum to which the ends of the generator's rotating conductor loops are connected (Figure 9.7). Each segment is separated from the others by a thin layer of insulation. For clarity, only one loop is shown in Figure 9.8(a) and (b). The commutator is shown as separated into two halves. The two ends of the loop are labelled as Conductor 1 and Conductor 2.

The output of the generator is connected to the commutator via carbon brushes, which are pressed against the commutator as it rotates. Generator rotation is achieved using mechanical energy provided by:

- Steam (generated by gas, oil, biomass or nuclear fission and fusion)
- A petrol, diesel or gas engine

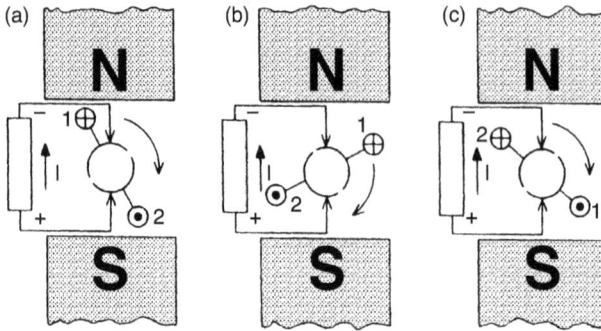

Figure 9.8   Principle of DC generator

- Wind
- Water flow
- By hand

In Figure 9.8(a), the loop approaches the point of maximum induced EMF. Current flows to the external load, which is connected to the loop via the commutator and brushes.

Conductor 1 – current flowing away from observer
Conductor 2 – current flowing towards observer

In Figure 9.8(b), the loop approaches the point of zero EMF. The brushes are close to the point where they will change connection from one commutator half to the other. Once that changeover has occurred conductors 1 and 2 will each cut the flux in the opposite direction to Figure 9.8(a).

In Figure 9.8(c), the changeover has occurred.

Conductor 1 – current flowing towards the observer
Conductor 2 – current flowing away from the observer

Note that each brush will always pick up either the positive or negative current, as required, because the commutator allows it to connect to whichever conductor is passing through the flux in a particular direction.

So, from Figure 9.8(a), (b) and (c), we can see that the commutator ensures that whichever conductor is passing the north pole, it is always connected to the negative end of the load, and the conductor passing the south pole to the positive end.

Practical DC generators have many conductor loops, and commutators with a large number of segments, so the output is steady. The magnetic flux cut by the loops on the rotating armature must be as great as possible to ensure maximum induced EMF, and this is arranged by means of a magnetic circuit. This circuit consists of

- iron polepieces
- the iron body of the rotor and the frame of the machine, which is called the yoke.

A DC machine must have an even number of poles (2, 4, 6, 8, etc.). The conductor loops in which the EMF is induced, referred to as the armature winding, are let into slots in the surface of an iron cylinder that is called the rotor. This is the rotating part of the machine.

The magnetic flux is set up by electromagnetic field windings placed on **salient poles**. Because these are electromagnets, they require their own source of electricity, or excitation. This may seem a little strange for a machine designed to *produce*, not *use* electrical power. For petrol and diesel generators the engine battery can be used to provide this excitation. For other types, a separate DC source has to be available. The power required to produce the necessary electromagnetic excitation is very small compared to the generator output.

Small DC generators, such as those used to power lights on a bicycle, are fitted with permanent magnets, which precludes the need for excitation.

> **KEY TERM: Salient Poles – These are magnetic poles that protrude from the surface of the rotor or the stator, from the which the magnetic field is produced**

9.6   Static induction

The induced EMF (Section 9.2) was due to movement. The conductor is rotated through a stationary magnetic field. EMF would also be induced if the conductor were stationary and the field were moved past it. However, EMF can be induced without any physical motion at all.

The key to this is the fact that induction is caused by a change in a magnetic field. The movement of a conductor in a generator takes it through areas of differing flux density. It is not the movement itself that induces voltage and current into the conductor. It is the change of flux strength it experiences.

Mutual inductance

Consider two coils of wire placed side by side (Figure 9.9), but not touching or in electrical contact with each other. The first coil is connected in series with a battery and a switch so that a current can be made to flow in it and can then be switched off. The second coil has a voltmeter connected to its ends.

If the switch in the circuit of the first coil is operated, the instrument connected to the second coil is seen to 'kick' and then return to zero. This happens each time the switch is turned on or off. The needle moves in a different direction at each operation. Figure 9.9 shows the reason for the induced EMF.

1. In Figure 9.9(a), the switch is off and the first coil sets up no magnetic flux.
2. When the switch is on (Figure 9.9(b)), the first coil sets up a magnetic flux, some of which passes through, or 'links with', the second coil.
3. Because there has been a change in the flux linking the second coil, EMF is induced into that second coil.

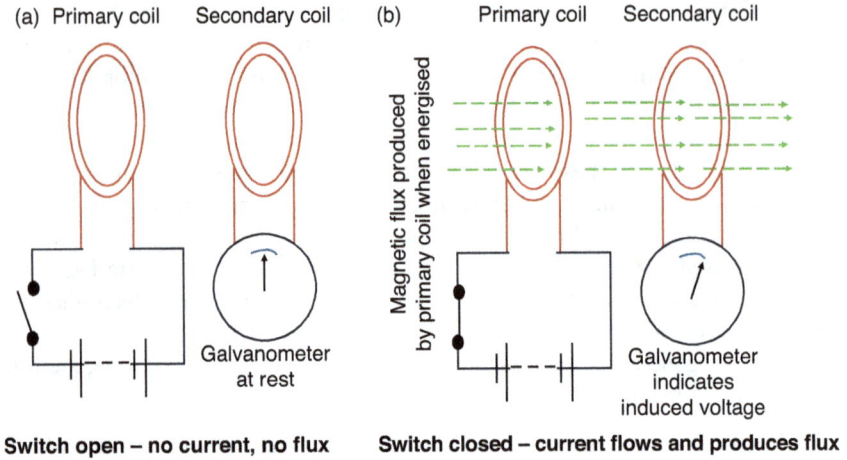

Figure 9.9   Mutual induction between two conductor coils

4. This EMF will collapse, however, unless there is another change. In other words, when the flux becomes steady, no EMF is induced.

The value of a statically induced EMF depends on the total magnetic flux change and the time it takes to complete this change.

$$e = \frac{\Phi}{t}$$

where e = induced EMF, V; Φ = total magnetic flux change, Wb; and t = time for flux change, s.

This, the EMF in volts induced at any instant of time is equal to the rate of change of magnetic flux at that instant. This is measured in webers per second.

Example 9.5

If a current change in a coil of 200 turns induces an average EMF of 25 V, what will be the total flux change if the current takes 50 ms to complete its change?

$$e = \frac{\Phi}{t} \text{ so } \Phi = et$$

Since the coil has 200 turns and the total induced EMF is 25 V, the induced EMF per turn will be

$$e = \frac{25}{200} = 0.125 \text{ V}$$

$$\Phi = et = 0.125 \times 50 \times 10^{-3} = 0.00625 \text{ Wb or } 6.25 \text{ mWb}$$

If the left-hand coil is fed from a source of alternating current, the magnetic flux it sets up will be continually changing. An alternating EMF will be induced in the right-hand coil. This is the principle of the transformer, which will be considered in Chapter 10.

Self-inductance

A further study of Figure 9.9 will show that a change of magnetic flux linkages has taken place in the left-hand coil, as well as in the right-hand coil. This induces a second EMF into the left-hand coil, but this EMF will oppose the battery voltage and try to slow the change of current.

   This self-induced EMF is sometimes called a back EMF. Any circuit that has the property of inducing such an EMF into itself is said to be self-inductive, or just inductive. All circuits are, to some extent, self-inductive, but some conductor arrangements give rise to a much greater self-inductance than others. The unit of self-inductance, which has the symbol L, is the henry (symbol H). The property of self-inductance is discussed fully in *Electrical Craft Principles, Volume 2.*

9.7   Summary of formulas for Chapter 9

$$e = Blv \quad B = \frac{e}{lv} \quad l = \frac{e}{Bv} \quad v = \frac{e}{Bl}$$

where e = induced EMF, V; B = flux density of magnetic field, T; l = length of conductor in the field, m; v = velocity (speed) of the conductor, m/s.

$$e = \frac{\Phi}{t} \quad \Phi = et \quad t = \frac{\Phi}{e}$$

where e = average induced EMF, V; Φ = total magnetic flux cut or total magnetic flux change, Wb; t = time taken to cut flux or time taken for flux change, s.

   Fleming's right-hand (generator) rule:

- First finger points in the direction of magnetic flux.
- Second finger points in the direction of induced EMF.
- Thumb points in the direction of conductor movement through the magnetic field

9.8   Exercises

1. Examine the diagrams in Figure 9.10 that show a conductor being moved in a magnetic field, and state
    (a) the direction of induced EMF for Figure 9.10(a)
    (b) the direction of induced EMF for Figure 9.10(b)
    (c) the magnetic polarity for Figure 9.10(c)
    (d) the direction of conductor movement for Figure 9.10(d)

(a)   (b)   (c)   (d)

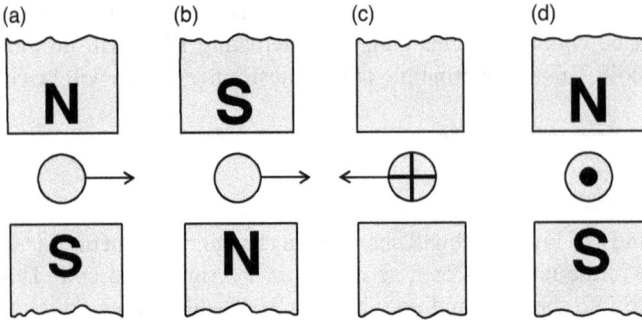

Figure 9.10   Diagrams for Exercise 1

2. A conductor is moved at a speed of 10 m/s directly across a magnetic field of flux density 15 mT, and has an EMF of 0.3 V induced in it. What is its effective length?

3. At what speed must a conductor of effective length 180 mm be moved at right angles to a magnetic field of flux density 0.6 T to induce in it an EMF of 0.324 V?

4. A conductor of effective length 200 mm connected across a milliammeter of resistance 5 Φ is moved through a magnetic field of flux density 0.5 T. If the milliammeter reads 40 mA, at what speed must the conductor be moving?

5. What EMF will be induced in a conductor of effective length 80 mm that is moving with a velocity of 15 m/s through a magnetic field of flux density 0.4 T?

6. One conductor of a generator is 500 mm long and moves at a uniform velocity of 20 m/s in the pole flux that as an average density of 0.4 T. What is the average EMF induced in the conductor? If the winding has 200 of these conductors connected in series, what is the total generated EMF?

7. A conductor is subjected to a magnetic flux changing at the rate of 4 Wb/s. What EMF is induced in the conductor?

8. An average EMF of 1.5 V in a conductor while the initial linking flux of 0.25 Wb is falling to zero. How long does the flux take to collapse?

9. A millivoltmeter connected to a conductor reads a steady 20 mV for 3 s while the conductor is subjected to a changing magnetic flux. Calculate the total flux change.

10. When a magnet is being inserted into a coil of wire, what factors govern
    (a) the direction
    (b) the magnitude of the induced EMF?

11. Describe with the aid of a sketch a simple loop generator that consists of a single wire loop rotating between the poles of permanent magnet. Show the output of the loop taken from slip rings, and sketch a graph of the output voltage to a base of time.

12. Draw a diagram to show a simple two-part commutator which can be substituted for the slip rings of the generator of Exercise 11. Describe how the commutator functions, and sketch a graph of the output from the machine.

13. Describe, with the aid of a sketch, the construction and action of a simple direct current generator. State
    (a) the factors on which the generated EMF depends
    (b) how the generated EMF can be controlled.

9.9   Multiple-choice exercises

9M1 The word 'induction' in the electrical sense is taken to mean
    (a) the ceremony that is held when someone joins an organisation
    (b) the production of EMF due to a change in linking magnetic flux
    (c) the amount of current flowing in a resistor when a voltage is applied
    (d) the production of magnetic flux when current flows in a coil

9M2 The value of EMF induced in a conductor when it moves through a magnetic field depends on
    (a) the thickness of the conductor
    (b) the total field flux and the length of the conductor subject to it
    (c) the shape of the magnetic field and the size of the conductor
    (d) the flux density of the magnetic field, the velocity of the conductor and the length of it in the field.

9M3 If a conductor wound in a loop is moved through a magnetic field
    (a) the induced EMF will be greater than for a single conductor
    (b) the device will be a kind of electric motor
    (c) the induced EMF will be smaller than for a single conductor
    (d) the individual EMFs in each loop will cancel to give no total EMF

9M4 If a conductor moving at a constant speed of 8 m/s through a magnetic field of flux density 0.65 T, has an EMF of 1.3 V induced, its length must be
    (a) 6.76 m
    (b) 0.106 m
    (c) 0.25 m
    (d) 0.148 m

9M5 A coil of 1000 turns has one side moved through a magnetic field of flux density 50 mT at a speed of 4.2 m/s. If 12 cm of each conductor is subjected to the flux, the total EMF induced in the coil will be
    (a) 25.2 V
    (b) 2.7 kV
    (c) 21 mV
    (d) 1.19 V

9M6 A coil of 450 turns carries current that sets up a total magnetic flux of 42 mWb. If an average EMF of 540 V is induced when the current is switched off, the current collapses to zero in
(a) 50.4 ms
(b) 3.5 s
(c) 28.6 ms
(d) 35 ms

9M7 The polarity of the magnet system shown in the diagram below must be
(a) both poles positive
(b) north pole at the top
(c) north pole at the bottom
(d) south pole at the top

9M8 Fleming's hand rule can be used to relate the directions of the magnetic field, the current and the motion, for induced EMF. This rule can only give the correct results if we use
(a) both hands
(b) the left hand
(c) the right hand
(d) either hand

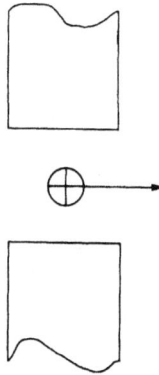

9M9 When using Fleming's hand rule the first finger always points the direction of the
(a) induced EMF
(b) magnetic field
(c) current flow
(d) force on the conductor

9M10 The EMF induced in a rectangular loop of wire that rotates in a magnetic field is
(a) alternating
(b) very low
(c) direct
(d) dangerous

9M11 When a conductor moves at high speed along a magnetic field so that is cuts no magnetic flux, the EMF induced in it will be
(a) zero
(b) a maximum
(c) alternating
(d) direct

9M12 The automatic changeover switch on the rotor of a DC machine is called the
(a) polepiece
(b) commuter
(c) brushes
(d) commutator

9M13 Static induction is the principle behind the operation of the
(a) generator
(b) motor
(c) transformer
(d) computer

Chapter 10

Transformers

10.1 Introduction

The transformer is undoubtedly the most important of all electrical machines. Transformers range in size from the miniature units used in some electronic applications to the huge transformers used in power stations. Although the methods of construction may differ widely, all transformers follow the same basic principles. Most transformers are required to provide an output voltage which is greater or less than that applied to the transformer.

10.2 Transformer construction

The basic components of a transformer are as follows.

Core

The core of a transformer is made from pure iron or a ferrous metal; the main four materials used are:

- Silicon steel – the most common because of its exceptional magnetic properties
- Amorphous alloys – produces energy-efficient cores
- Ferrite core – a lightweight core material
- Nanocrystalline core – used for high-frequency transformers

The purpose of the core is to create a strong electromagnetic field. Because the flux must be in a constant state of change to effect electromagnetic induction, the transformer will be supplied with AC.

The core has a hollow centre to enable windings to be installed. Generally, the core will be a four-sided shape, but in certain cases, such as in a residual current device (see Chapter 16), it can be circular, shaped like a doughnut. This is called a toroidal core, which is more power efficient because its shape lends itself to an even distribution of windings.

Windings

Sets of insulated conductors are wound around the iron core. These are called windings. These are:

- Primary – supply side of the transformer
- Secondary – output or load side of the transformer

The symbol for the number of turns on a transformer winding is N. The more turns on the output side, the higher its induced voltage will be.

Transformer windings are made from copper or aluminium and can be in wire or flat tape form. They can be insulated with:

- Epoxy resin
- Electrical grade paper
- Pressboard
- Wood and insulated wood
- Insulating tape
- Phenolic laminated paper base sheet
- Phenolic laminated cotton fabric sheet

Cooling system

Heat is a major problem with transformers. For the large supply transmission and distribution types, overheating can be catastrophic and result in explosion and fire. The three main cooling systems for transformers are:

- Oil – a mineral oil made up of several compounds that also improve insulation and reduce corrosion
- Gas – mainly sulphur hexafluoride (SF6) and nitrogen
- Air – used to cool smaller transformers. The small size of a charging plug, for example, means that it is surrounded by enough air to keep it cool. Other equipment uses fans to draw air over internal transformers and associated components.

Convection

Oil- and gas-filled transformers use convection to maintain a steady temperature. The hot oil or gas is forced out through the pipes or fins on the sides of the transformer, where it cools before returning to the transformer core. This is a natural current that runs through liquid and gas when heated. An example of cooling fins can be seen on the right-hand transformer in Figure 10.1.

Buchholz relay

The Buchholz relay is a safety device used for large oil-filled transformers. It monitors oil levels and temperature and also detects changes in the chemical make-up of the medium. If these changes are severe enough to pose a danger, the relay will cut off the transformer's supply (Figure 10.2).

Figure 10.1   *Examples of pole-mounted transformers. These will be used to transform 11 kV down to 400/230 V. Note cooling fins on right hand image.*

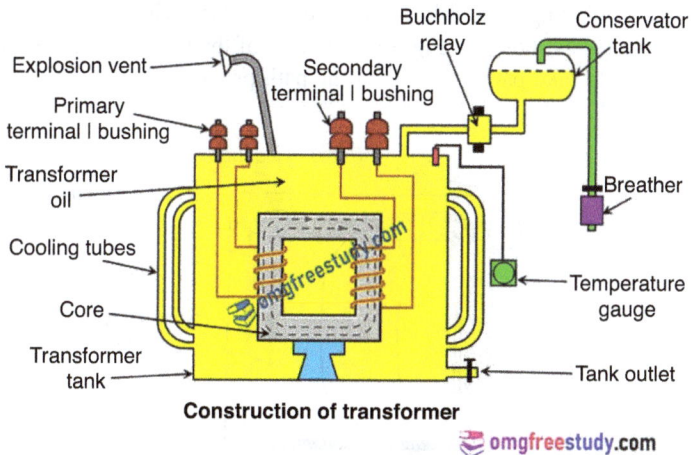

Construction of transformer

Figure 10.2   *Construction of a transformer showing main components (by kind permission of OMGFreestudy.com)*

10.3   Operating principles

Alternating current (AC) is used for transformer supply because it causes constant change in primary winding current and resulting magnetic flux. The EMF induced in the secondary winding will also be AC.

Two windings (primary and secondary) are arranged about the core.

1. When voltage is applied to the primary winding, magnetic flux forms in the core.
2. Because the supply is AC and, therefore, constantly changing, the strength of the core magnetic flux will also change.
3. This change induces EMF into the secondary winding.

Figure 10.3 shows the simple principle of the transformer. It should be noted that there is no electrical connection between the two windings. The link is magnetic only. Because of this, BS 7671 requires one point on the secondary winding of most transformers to be earthed to prevent a high potential between that winding and earth.

An EMF is induced in the secondary winding only while the current (and therefore flux) in the primary is changing. For this reason, transformers are ideally used on AC supplies. They can sometimes be used from a DC supply that is rapidly switched on and off so that the current continually rises and falls.

To sum up:

- AC supply to the primary winding sets up a constantly changing magnetic flux in the core
- Changing flux induces current into secondary winding

For a transformer to be as efficient as possible, all the magnetic flux produced by the primary winding should link with the secondary winding. If a transformer were made exactly as shown in Figure 10.4, a lot of the magnetic flux produced by the primary would take paths other than the path provided by the iron core.

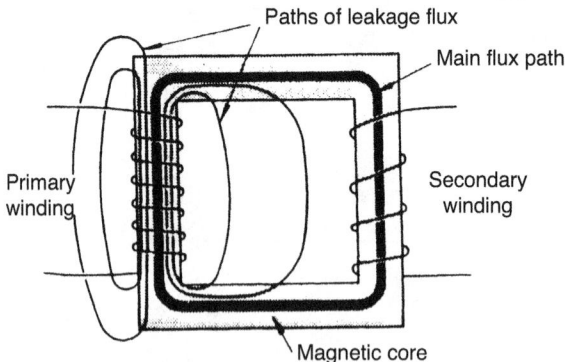

Figure 10.3   Simplified arrangement for transformer showing main and leakage flux paths

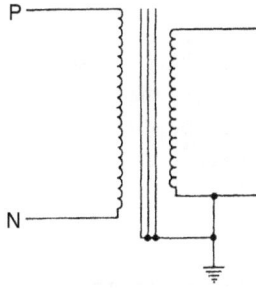

Figure 10.4 Circuit diagram symbol for double-wound transformer

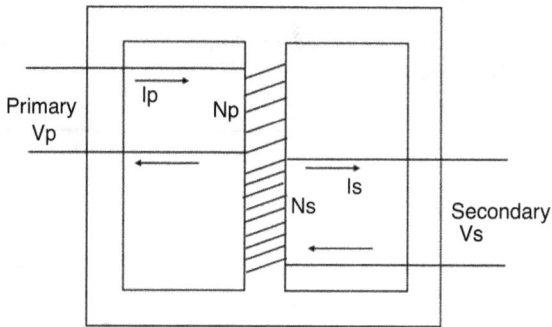

Figure 10.5 Core-type transformer

To reduce this leakage flux, primary and secondary windings may be split into sections, half of each winding being placed on each side of the core. Primary and secondary may also be wound one over the other. For power transformers, the core-type construction shown in Figure 10.5 is often chosen. The windings are placed on the centre limb. The side limbs provide a path for return flux.

The EMF induced in each turn of the transformer secondary winding will depend on the rate of change in the magnetic flux that links with it.

- Since the turns of a winding are all in series with each other, the total EMF induced in a winding will be the product of the EMF per winding turn and the number of turns.
- The total induced EMF is proportional to the number of turns, and the ratio of EMFs in the two windings will be the same as the ratio of their numbers of turns; that is,

$$\frac{Ep}{Es} = \frac{Np}{Ns}$$

where E_p = EMF induced in the primary, V; E_s = EMF induced in the secondary, V; N_p = number of primary turns; N_s = number of secondary turns.

Transformer ratios

In practice, both windings will have resistance, so that the voltage across the windings on load will not be quite the same as the EMFs induced in them. The differences are the result of voltage drops in the windings. Power-transformer windings have low resistance, and only a small error will be introduced by assuming that the terminal voltage is the same as the induced EMF, so that

$Ep = Up$ and $Es = Us$

Example 10.1
A transformer is wound with 480 turns on the primary and 24 turns on the secondary. What will be the secondary voltage if the primary is fed at 230 V?

$$\frac{Up}{Us} = \frac{Np}{Ns} \quad Us = \frac{NsUp}{Np} \quad Us = \frac{230 \times 24}{480} = 11.5 \text{ V}$$

Example 10.2
A transformer with a turns ratio of 2:9 is fed from a 230 V supply. What is its output voltage?

Transformer ratios are always given in the form 'primary:secondary'. This transformer has two turns on the primary for every nine on the secondary, so $Np/Ns = 2/9$.

$$Us = Up \times \frac{Ns}{Np} = 230 \times \frac{9}{2} = 1035 \text{ V}$$

If the secondary voltage is greater than the primary, the transformer is referred to as **step-up**, and if smaller as **step-down**.

If we neglect losses, the input and output powers for a transformer will be the same. Put simply,

$UpIp = UsIs$

Transposing and adding the turns ratio to the expression, we have

$$\frac{Up}{Us} = \frac{Is}{Ip} = \frac{Np}{Ns}$$

Current is inversely proportional to voltage, so if a transformer steps up the voltage from primary to secondary, the current will be less in the secondary than in the primary.

Example 10.3
A transformer has a turns ratio of 5:1 and is supplied at 230 V when the primary current is 2 A. Calculate the secondary current and voltage

$$Us = \frac{UpNs}{Np} = \frac{230 \times 1}{5} = 46 \text{ V}$$

$$Is = \frac{UpIp}{Us} = \frac{230 \times 2}{46} = 9.58 \text{ A}$$

10.4 Transformer losses

The transformer iron core will heat up owing to what are called 'core losses'. These losses are subdivided into the following.

Hysteresis losses

This subject will be considered more fully in *Electrical Craft Principles, Volume 2*. Hysteresis losses can never be removed, but can be reduced by the addition of a small percentage of silicon into the magnetic iron core.

Eddy-current losses

The iron core is itself a conductor, into which an EMF is induced by the transformer's changing magnetic flux. The EMF causes a current to eddy back and forth in the iron. This unwanted current causes wasted energy in the form of heat.

To reduce eddy currents to a minimum, the core of a transformer is subdivided into small parts; each part has only a small share of the total induced EMF and has high resistance owing to its small cross-section, so that eddy currents are small.

In power transformers, this subdivision is carried out by building up the core with layers of thin plates, called laminations. Each lamination is insulated from its neighbours by being varnished or oxidised or covered with paper on one side. The laminations must be arranged so that the magnetic flux is set up along them, and not across the insulation between them, which is non-magnetic and would reduce the strength of the field (Figure 10.6).

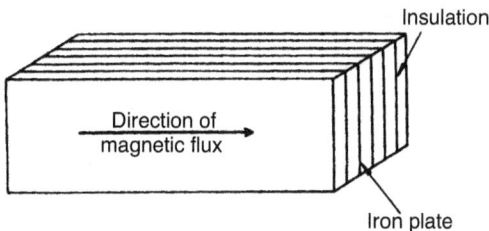

Figure 10.6 Part of magnetic circuit build-up of laminations (lamination thickness has been exaggerated)

A winding machine is used to make the transformer windings, which are then slipped over the core. A core of the type shown in Figure 10.4 would be made up of E and I, or of U and T-shaped sections. Joints in the core must fit as well as possible to allow the maximum magnetic flux to be set up. Laminations are interleaved at the corners to improve the effective fit.

Copper (I^2R) losses

Copper losses are due to the heating effect of the primary and secondary currents passing through their respective windings. Although the resistance of these windings is kept as low as possible, for power transformers it cannot be zero.

Copper losses can only be prevented if the current is reduced to zero. In practice, if the secondary winding of a transformer is open-circuited (so that it carries no current), the primary winding will still carry a small current. This current provides the ampere-turns necessary to set up magnetic flux in the core. However, this magnetising current is quite small, so, in this case, the copper loss due to it may be ignored. The total copper losses of a power transformer on no load may be assumed to be zero.

Power loss in a resistive circuit is given by the expression $P = I^2R$, and since winding resistances are largely constant, copper losses depend on the square of the load current. This means that a transformer operating on half load will have only one quarter of the copper loss it has when on full load.

$$\text{Copper loss } Pc = \left(\frac{\text{actual load}}{\text{full load}}\right)^2 \times \text{full load copper loss}$$

The following measure can be taken to reduce copper losses in a transformer:

- Windings with a larger cross-sectional area
- Enhanced cooling
- Reduce load current
- Optimised windings – for example, using parallel conductors

10.5   Summary of formulas for Chapter 10

Voltage, current and turns ratios

$$\frac{Up}{Us} = \frac{Np}{Ns} \quad Up = \frac{UsNp}{Ns} \quad Us = \frac{UpNs}{Np} \quad Np = \frac{UpNs}{Us} \quad Ns = \frac{UsNp}{Up}$$

$$\frac{Up}{Us} = \frac{Is}{Ip} \quad Up = \frac{UsIs}{Ip} \quad Us = \frac{UpIp}{Is} \quad Ip = \frac{UsIs}{Up} \quad Is = \frac{UpIp}{Us}$$

Copper losses

$$P = I^2R \quad \text{Copper loss } Pc = \left(\frac{\text{actual load}}{\text{full load}}\right)^2 \times \text{full load copper loss}$$

10.6 Exercises

1. Make a labelled sketch of a square cross-section core for a single-phase transformer showing its detailed construction. Describe briefly its various components and the precautions to be taken during its assembly.
2. A transformer connected to a 230 V supply has 600 primary turns, and is required to have a secondary voltage of 40 V. How many turns must it be wound on the secondary.
3. A transformer with a turns ratio of 500:40 has an output of 10 V. What voltage is supplied to the primary winding.
4. A transformer primary winding connected across a 400 V supply has 600 turns. How many turns must be wound as the secondary if an output of 1328 V is required?
5. Make sketches indicating current flow and resulting magnetic field when current flows in a solenoid with a steel core in the shape of a closed ring. Explain how this construction can form the basis of a transformer if supplied with alternating current.
6. Draw a sketch of a transformer showing core and coils and mark in the flux path.
7. Explain how the transformer works.
8. A transformer is supplied at 230 V. The primary winding has 4800 turns, and takes 1 A for every 10 A delivered by the secondary. Calculate
   (a) the number of turns on the secondary
   (b) the secondary voltage.

9. Describe, with the aid of a sketch, a simple step-up transformer and explain its action.
10. A transformer has primary and secondary voltages of 720 V and 300 V, respectively. What will be the primary current when the secondary delivers 10 A?
11. A transformer has 1500 primary turns and 75 secondary turns. What current will the secondary provide if the primary carries 5 A?
12. A transformer has 200 primary turns, and is fed at 120 V. If primary and secondary currents are measured at 1.5 A and 120 mA, respectively, calculate the number of secondary turns and the secondary voltage.

10.7 Multiple-choice exercises

10M1 The two windings on a simple transformer are called the
   (a) magnetic circuits
   (b) first and second windings
   (c) primary and secondary
   (d) connections

10M2 The magnetic flux produced by one transformer winding that does not link with the other is called
(a) linking flux
(b) leakage flux
(c) main flux
(d) useful flux

10M3 For a transformer
(a) turns ratio is equal to the voltage ratio
(b) voltages in both windings are equal
(c) turns ratio is equal to the current ratio
(d) primary volts per turn are twice secondary volts per turn

10M4 The voltage ratio of a transformer is 230 V:15 V. If the primary current is 250 mA, the secondary current will be
(a) 15.6 mA
(b) 250 mA
(c) 3.83 A
(d) 400 mA

10M5 If the secondary voltage of a transformer is greater than its primary voltage, it is called a:
(a) dangerous transformer
(b) step-up transformer
(c) step-down transformer
(d) voltage transformer

10M6 Eddy-current losses in a transformer are reduced by making the core using
(a) laminations
(b) solid magnetic steel
(c) high-hysteresis-loss material
(d) two windings

Chapter 11

Electrical motor principles

11.1   Introduction

Electric motors have a vast range of applications, from the large powerful engines that drive fully laden electric trains to the tiny motor that drives the hands of a digital watch. Every electric pump, lathe, power drill, washing machine, hedge cutter and electric car will have such a motor at its heart (Figure 11.1).

In Chapter 9, it was explained that an electric current gives rise to a magnetic field. If a current-carrying conductor is placed in a second, external magnetic field, the two magnetic fields react and the conductor is subjected to a mechanical force. This force can be used to rotate an armature, and the result is an electric motor.

It is interesting to reflect on how much our civilisation depends on the electromagnetic principles of the generator and the motor. Without these machines, the world would be a very different place.

11.2   Force on a current-carrying conductor in a magnetic field

A current-carrying conductor experiences a force when it lies at right angles to a magnetic field. We can verify this natural law by using a piece of wire, a permanent magnet and a battery. If the wire is placed between the magnet's poles and its ends momentarily connected across the battery terminals, it will jump from its position.

Figure 11.2 explains why this force occurs. Figure 11.2(a) shows the magnetic field around a live conductor. The conductor is drawn in cross-section and is carrying a current away from us. Figure 11.2(b) shows the field set up by the magnetic poles in which a conductor is placed. No force is exerted on the conductor when it carries no current. In Figure 11.2(c), the conductor now carries a current that flows away from the observer. A spiral magnetic field forms about the conductor when it carries current. The flux direction is clockwise with the direction of current flow. At the point where the conductor's flux is oriented in the same direction as that of the magnetic field, repulsion will occur. Attraction will occur at the point where the conductor's flux opposes that of the magnetic field.

It is also true that since lines of magnetic flux never cross, the two fields cannot exist simultaneously in their individual forms, and the resultant field takes up the shape shown in Figure 11.2(c).

Figure 11.1   Electric motor. Image credit: VEM motors GmbH, via Wikimedia Commons, CC BY-SA 4.0.

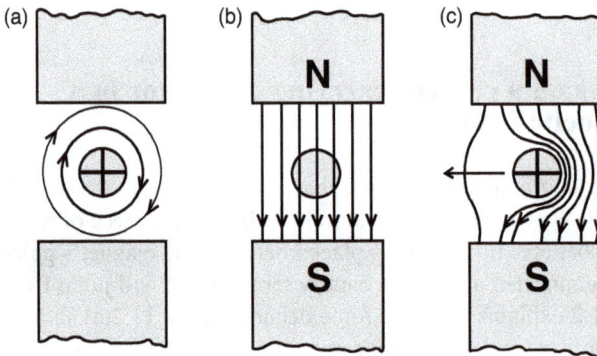

Figure 11.2   Force on current-carrying conductor lying in magnetic field

The stronger field to the right of the conductor tries to contract and exerts a force on the conductor in much the same way as a stone in a catapult. If free to do so, the conductor will move to the left. If the conductor is moved out of the influence of the magnetic field, the force on the conductor will weaken and dissipate.

If either the polarity of the magnet or the current in the conductor is reversed, the direction of force on the conductor will be reversed. If both magnetic and current polarity are reversed, the force remains in the same direction. We can calculate the force on a current-carrying conductor in a magnetic field, provided the conductor is at right angles to the field:

$$F = BIl$$

where F = force on conductor, N; B = flux density of magnetic field, T; l = length of conductor in the field; and I = current flowing in conductor, A.

Example 11.1
A conductor, 0.2 m long, carries a current of 25 A at right angles to a magnetic field of flux density 1.2 T. Calculate the force exerted on the conductor.

$$F = BIl = 1.2 \times 0.2 \times 25 = 6 \text{ N}$$

Example 11.2
How much current must a conductor of an electric motor carry if it is 900 mm long and is situated at right angles to a magnetic field of flux density 0.8 T, if it has a force of 144 N exerted on it?

$$F = BIl; \quad \text{therefore, } l = \frac{F}{Bl} = \frac{144}{0.8 \times 0.9} = 200 \text{ A}$$

11.3   Relative directions of current, force and magnetic flux

It is important to know the direction of the force on a current-carrying conductor in a magnetic field because, in practice, a motor shaft needs to rotate in the direction required by the machinery it drives.

One method is to draw out the field as shown above. Fleming's left-hand (motor) rule, which indicates the directions of the current, field and force, can also be used.

The thumb, the first finger and the second finger of the left hand are extended so that all three are at right angles to each other (Figure 11.3).

- First finger points in the direction taken by the magnetic field.
- SeCond finger in the direction of current flow.
- ThuMb gives the direction of motion of the conductor as a result of the force applied to it.

A little practice will show how easy this rule is to apply. It must be carried out using the left hand, and applies only to the motor effect.
Remember:

- Motors = Fleming's **Left** Hand Rule.
- Generators = Fleming's **Right** Hand Rule (Gener-RIGHTER).

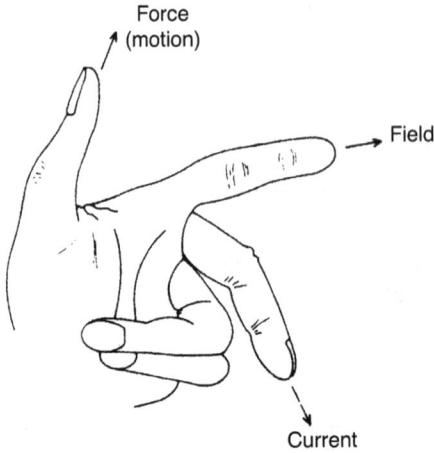

Figure 11.3   Position of left hand for application of Fleming's left-hand rule

---

Example 11.3
Refer to Figure 11.4 and give

(a) the direction of the force on the conductor in Figure 11.4(a)
(b) the polarity of the field system in Figure 11.4(b)
(c) the direction of the current in Figure 11.4(c)
(d) the direction of the force on the conductor in Figure 11.4(d)

Applying Fleming's left-hand rule, or sketching the magnetic field shapes, gives the results

(a) right to left
(b) north pole at the top
(c) out of the paper
(d) left to right.

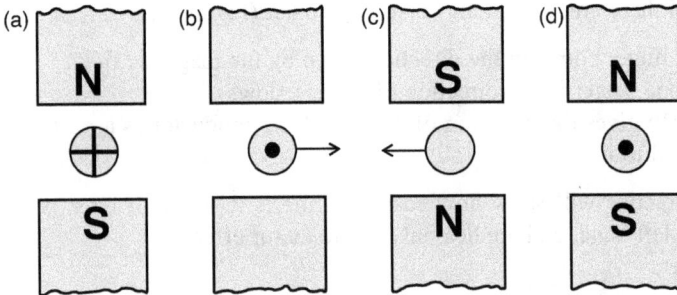

Figure 11.4   Figures for Example 11.3

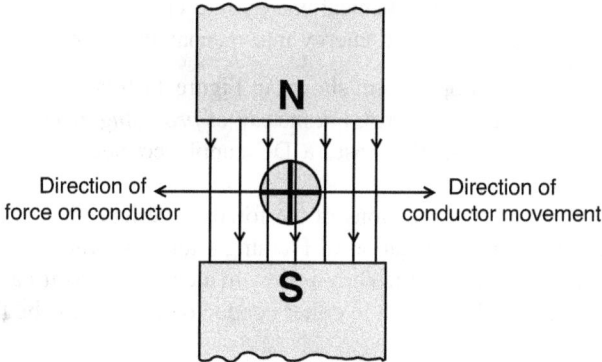

Figure 11.5 Differing directions of conductor movement and force (Lenz's law)

11.4   Lenz's law

In simple terms, Lenz's Law states that a current in a conductor will always oppose the change that produced it (Figure 11.5).

- When a magnetic field changes, the induced current will create its own, separate magnetic field.
- The direction of this separate field will counteract that of the changed field.
- This is why the current arising from a back EMF in a winding (see Chapter 9) opposes the supply current.

In the case shown, extra energy must be used to overcome the reverse force. The work needed to overcome it will increase as the current increases, so that for a generator we have to put more mechanical energy in to achieve the required electrical energy output.

The reverse force will not, of course, completely stop the conductor. If it did so, the induced EMF, and hence the current producing the reverse force, would disappear.

The law also affects induction in a circuit, which is due to a change in linking magnetic flux (Chapter 9). The EMF induced by the changing current will always be in such a direction as to resist that change.

- If the current is reducing, the back EMF will be in the same direction as the current and will try to maintain it.
- If the current is increasing, the back EMF will oppose it and try to prevent the increase.

11.5   Direct-current motor principles

The direct-current motor is basically the same as the direct-current generator, which was considered in Chapter 9. Both machines are energy converters.

- The generator converts mechanical energy into electrical energy.
- The motor converts electrical energy into mechanical energy.

The simple rectangular loop system shown in Figure 11.6 is the same as the generator arrangement of Figure 9.7(a), but instead of *providing* electricity, it must be *supplied with* electricity, in this case, a DC supply connected to the motor via brushes and a commutator.

Figure 11.7 shows the directions of the forces experienced by the conductors, which can be verified by application of Fleming's left-hand rule.

The commutator reverses the current flow in a conductor as it passes from one pole to the next so that the current in either conductor will always be the same as it

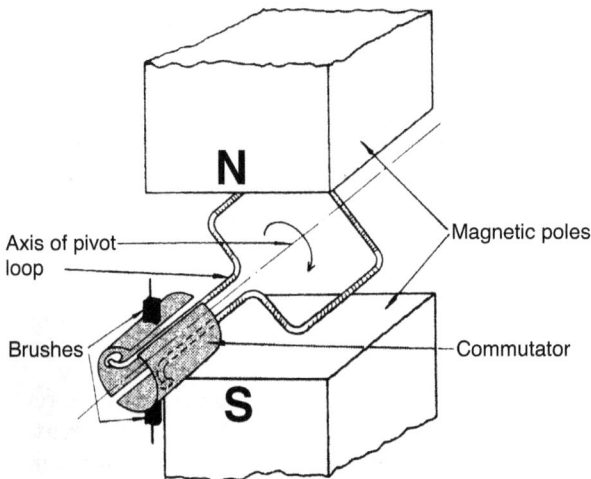

Figure 11.6   Loop connected to simple commutator and able to rotate in magnetic field

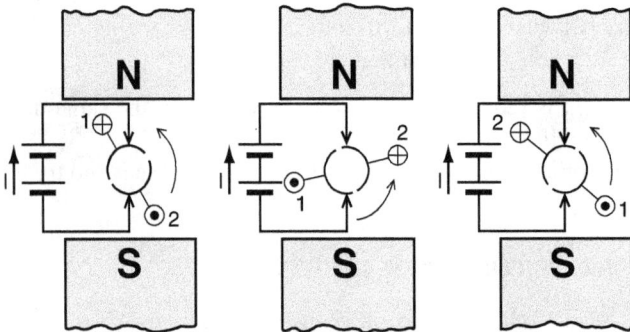

Figure 11.7   Principle of DC motor

passes a given pole. The direction of the force will therefore be the same, and the loop will rotate continuously in a given direction.

At the instant when the brushes are passing over the joints in the commutator, the conductors will be moving *along* the lines of magnetic flux and will experience no force. In practice, the speed of rotation of the loop will keep it moving until this 'dead spot' is passed. Like the generator, the practical DC motor has many loops and a multi-segment commutator. As a result, the force on the machine is nearly constant, and no 'dead spot' occurs. The construction of the DC machine was described briefly in Chapter 9.

11.6   Moving-coil instrument

Although this chapter subject is concerned with the basics of electric motor operation, the same electromagnetic principles apply to moving coil test instruments.

In Section 11.2, we saw that the force exerted on a conductor carrying current in a magnetic field ($F = BIl$) depends on:

- magnetic field strength
- length of conductor in the field
- conductor current

If field strength is made constant by the use of a permanent magnet, and the conductor is in the form of a coil of fixed length, the force must depend only on the current. Using this principle, an instrument can be made to measure the current it carries, giving a deflection depending on the force exerted on its coil and, therefore, on its current. This instrument is called the permanent-magnet moving-coil instrument.

It resembles a miniature DC motor, but instead of having a commutator to allow continuous rotation, current is fed into the coil through hair springs which limit the angle of rotation.

Although most electrical measurement instruments are digital, moving coil versions are still in use where a quick visual indication is helpful. The movement of a needle or pointer is sometimes easier to interpret than a set of figures.

In many types, the shape of the permanent magnet is similar to that shown in Figure 11.8, the magnetic circuit being completed with shaped soft-iron pole shoes and cylindrical core so that a radial and uniform magnetic field is set up by the airgap. The components of the moving coil instrument are as follows:

- The coil of fine insulated wire wound on an aluminium former is pivoted to swing in this field, which it will always cut at right angles (Figure 11.9).
- Two phosphor-bronze hairsprings make electrical connections to the moving coil. They also:
  o   Limit the coil swing (controlling torque).
  o   Return the movement to the zero position when no current flows (restoring torque).

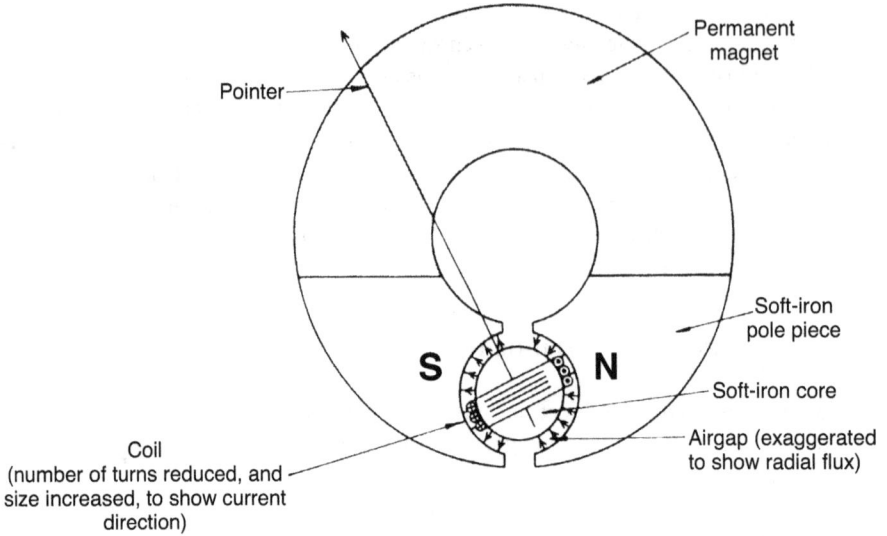

Figure 11.8   General arrangement of permanent-magnet moving-coil instrument

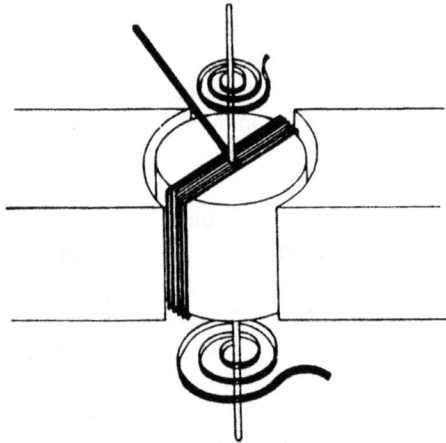

*Figure 11.9   View of permanent-magnet moving-coil instrument, showing coil,
pointer, spindle and springs*

- Coil movement occurs when current flows in its windings. It becomes, in effect, a series of current-carrying conductors lying in a magnetic field.
- The force on the two sides of the coil turns it against the torque of the control springs. Equilibrium occurs when, due to the current, the deflecting torque is equal and opposite to the controlling torque produced by the hairsprings.

- A light aluminium pointer is fixed to the coil and moves over a scale to measure the current.

Damping

With most instruments, the movement need comes to rest quickly at its reading, without excessive oscillation. This provision is known as damping. With the permanent-magnet, moving-coil instrument, damping is achieved automatically. The EMF induced in the aluminium-coil former as it swings in the magnetic field, causes an eddy current to flow. In accordance with Lenz's Law, this produces a damping force which opposes the movement producing it.

Advantages and disadvantages

In their standard form, these instruments are limited to an angular movement of about 120°, but special movements are available to give circular scales. In these instruments, the needle is capable of swinging through an angle of about 300°. This increased scale length gives an improvement in the accuracy of reading. Alternatively, the same scale length can be accommodated on an instrument that takes up less space.

Since the torque and instrument deflection are proportional to current, the scale is linear; that is, there are equal spaces between equal divisions. Other advantages of this instrument are its

- Accuracy
- Sensitivity to small currents
- Ease with which it can be adapted for almost any value of current or voltage.

The main disadvantage of the instrument is its

- Inability to read values of AC at power frequencies. These currents give rise to an alternating torque.
- Movement, the instrument is unable to adjust to rapid variations set up by AC and remains at the zero position.
- The delicate moving system is easily damaged by rough handling, and the fine coil will not withstand prolonged overloading.

11.7   Summary of formulae for Chapter 11

For a conductor lying at right angles to a magnetic field,

$$F = BIl \quad B = \frac{F}{Il} \quad l = \frac{F}{BI} \quad I = \frac{F}{Bl}$$

where F = force on conductor, N; B = flux density of magnetic field, T; l = length of conductor in magnetic field, m; and I = current carried by conductor, A.

11.8   Exercises

1. A force of 10 N is exerted on a conductor 1.5 m long when carrying a current and lying at right angles to a magnetic field of flux density 1.5 T. What is the current?

2. If a conductor lying at right angles to a magnetic field of flux density 0.12 T experiences a force of 8 N when carrying a current of 5 A, what is the effective length of the conductor in the magnetic field?

3. Twenty millimetres of a conductor carrying a current of 15 mA is situated at right angles to a magnetic field, and experiences a force of 0.33 mN. What is the field flux density?

4. With the aid of a sketch, show how a force is produced on a current-carrying conductor lying in a magnetic field. Directions of the current, magnetic field and force should be shown.

5. What force is experienced by a busbar 2 m long which, under fault conditions, carries a current of 20,000 A in a magnetic field of flux density 100 mT?

6. One conductor on the coil of a moving-coil instrument is 10 mm long, and experiences a force of 2 µN when carrying a certain current. The airgap flux density is 0.8 T. What is the coil current?

7. Figure 11.10 gives examples of a current-carrying conductor lying in a magnetic field. State
    (a) the direction of the force on the conductor in Figure 11.10(a)
    (b) the direction of the force on the conductor in Figure 11.10(b)
    (c) the direction of the current in Figure 11.10(c)
    (d) the polarity of the magnet in Figure 11.10(d)

8. A moving-coil loudspeaker is required is provide a force of 0.2 N when its coil, which has an effective conductor length of 15 m, carries a current of 12 mA. What flux density must be set up by the magnet?

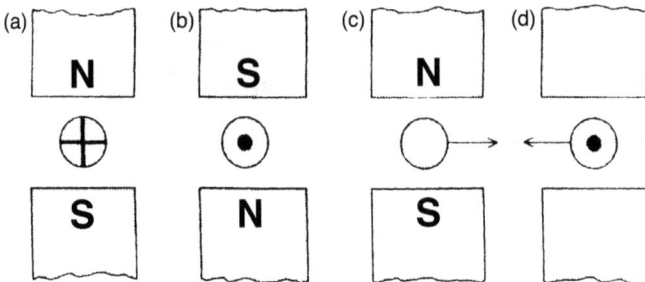

Figure 11.10   Diagrams for Exercise 7

9. What length of conductor must be present in the armature of a motor which has a pole flux density of 0.6 T and is required to provide a turning force of 300 N when the conductor current is 25 A?

10. (a) Consider a single conductor carrying a direct current and lying in the magnetic field between the poles of a two-pole DC motor. Make a sketch or diagram illustrating this. Assuming your own directions of magnetic field and current, indicate clearly the direction in which the conductor will tend to move. (b) A conductor 0.3 m long lies at right angle to a magnetic field of intensity 1.6 T and carries a current of 25 A. Calculate the force on the conductor.

11. Sketch a two-pole DC motor showing armature, poles, commutator and brushes, and explain how a continuous torque in the same direction is obtained.

12. Sketch a moving-coil instrument and label the main parts. State two disadvantages of this type of instrument.

13. Explain why a moving-coil instrument is unsuitable for use in AC circuits.

11.9 Multiple-choice exercises

11M1 A current-carrying conductor which is subjected to a magnetic field
   (a) must be PVC insulated
   (b) will corrode more quickly than if there is no field
   (c) will experience a force
   (d) will have an EMF induced in it

11M2 The formula relating the force on a conductor (F), its length (l), the current it carries (I) and the flux density of the magnetic field to which it is subject (B) is
   (a) $B = FlI$
   (b) $F = BIl$
   (c) $I = BFl$
   (d) $F = BlI$

11M3 A conductor experiences a force of 2.5 N when 45 cm of it is subjected to a magnetic field and it is carrying a current of 65 A. The magnetic flux density of the field is
   (a) 85.5 mT
   (b) 73.1 T
   (c) 0.855 mT
   (d) 0.855 T

11M4 The figure below shows a conductor lying in a magnetic field when carrying current. It will experience a force
   (a) into the paper
   (b) from left to right
   (c) towards the north pole
   (d) from right to left

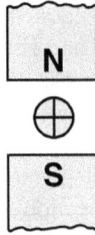

11M5 The rule relating the direction of current, magnetic field and force on a
conductor is
(a) Lenz's law
(b) Fleming's left-hand rule
(c) Fleming's right hand rule
(d) Ohm's law

11M6 Lenz's law states that
(a) a conductor carrying current experiences force when subject to magnetic field
(b) a commutator is necessary for a DC machine to work properly
(c) an induced EMF will always oppose the effect producing it
(d) when a conductor cuts a magnetic field an EMF will be induced in it

11M7 The simple electric motor shown in the figure below will
(a) rotate counter-clockwise
(b) not rotate at all
(c) rotate clockwise
(d) burn out

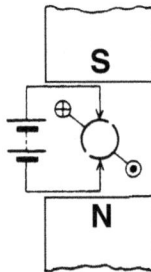

11M8   The major disadvantage of the permanent-magnet moving-coil instrument
is its
(a) inaccuracy
(b) nonlinear scale
(c) inability to measure alternating currents
(d) very heavy moving system

11M9   The coil of a permanent-magnet moving-coil instrument is wound with
very fine wire so that
(a) it can carry heavy current
(b) it is very light in weight
(c) it can easily have its range extended
(d) it will measure alternating and direct currents

11M10   The scale of a permanent-magnet moving-coil instrument is
(a) very cramped at the lower end
(b) difficult to read with accuracy
(c) reversed
(d) linear

Chapter 12

Basic alternating-current theory

12.1 What is alternating current?

Although direct current (DC) is in wide use in both the home and the commercial environment, the electricity supply is alternating current (AC). The easiest method of portraying an alternating quantity is to draw a graph showing how it varies with time, as in Figure 12.1.

- Any part of the graph that lies above the horizontal (or zero) axis represents current or voltage in one direction.
- Values below it represent current or voltage in the other direction.

The pattern given by the graph is known as the waveform of the AC system, and this usually repeats itself. There is no need for the waveform above the zero axis to have the same shape as that below it, although this is the case in most mains AC systems.

The waveform graph shows that AC rises in one direction to a maximum value before falling to zero and repeating in the opposite direction. Instead of drifting steadily in one direction, the electrons that form the current move backwards and forwards in the conductor. The time taken for an alternating quantity to complete its pattern (to flow in both directions and then return to zero) is called the periodic time (symbol T). This completed pattern is the result of one rotation of the generator, or cycle. The complete cycle is split into the

- positive half-cycle above the axis
- negative half-cycle below it

The number of complete cycles traced out in a given time is called the frequency (symbol f), usually expressed in hertz (Hz), which are cycles per second (c/s). If there are f cycles in one second, each cycle takes $1/f$ seconds, so that:

$$T = \frac{1}{f} \text{ and } f = \frac{1}{T}$$

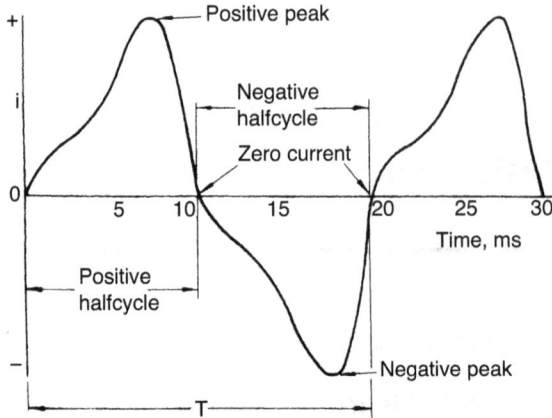

Figure 12.1   Graph of alternating current

Example 12.1
Calculate the frequency of the AC system shown in Figure 12.1.
From Figure 12.1, $T = 20$ ms $= 0.02$ s; therefore

$$f = \frac{1}{T} = \frac{1}{0.02} = 50 \text{ Hz}$$

A frequency of 50 Hz is the standard for the supply system in many parts of the world, including the UK, but 60 Hz systems are also common.

12.2   Advantages of AC systems

There are certain complications that occur when using AC supplies, which are absent with DC supplies. These complications are explained later in this chapter. However, the advantages of AC supplies have led to their general use. Some of the more important are as follows:

- An AC generator is more robust, less expensive, requires less maintenance, and can deliver higher voltages than its DC counterpart.
- The power loss in a transmission line depends on the square of the current carried ($P = I^2 R$). If the voltage used is increased, the current is decreased, and losses can be made very small.
- The simplest way of stepping up the voltage at the sending end of a line and stepping it down again at the receiving end is to use transformers, which will only operate efficiently from AC supplies.

- Three-phase AC induction motors are cheap, robust and easily maintained.
- Energy meters, to record the amount of electrical energy used, are much simpler for AC supplies than for DC supplies.
- DC systems are subject to severe corrosion, which is hardly present with AC supplies.

12.3 Values for AC supplies

The alternating current or voltage changes continuously, so that it is not possible to state its value in the same simple terms that can be used for DC.

Instantaneous values

These are the values at particular instants of time, and will be different for different instants. Symbols for instantaneous values are lower case:

v for voltage
i for current

Maximum or peak values

Usually occurring once in each half-cycle, maximum values are the greatest values reached during alternation. They are indicated by U_m for voltage and I_m for current.

Mean values

If a mean value is found over a full cycle, the positive and negative half-cycles will cancel out to give a zero result if they are identical. In such cases it is customary to take the mean value over a half-cycle. The mean value of this kind of waveform can be calculated as shown in Example 12.2. Symbols used are U_{av} for voltage and I_{av} for current.

Example 12.2
Table 12.1 gives the waveform of a half-cycle of alternating voltage. Find the frequency of the supply, its instantaneous values after 8 ms and 2.4 ms, the

Table 12.1 Waveform for Example 12.2

Time, ms	0	0.25	0.5	0.75	1.0	1.25	1.5	1.75	2.0
Volts, V	0	45	72	91	104	118	142	185	240

Time, ms	2.25	2.5	2.75	3.0	3.25	3.5	3.75	4.0
Volts, V	278	295	300	280	248	195	85	0

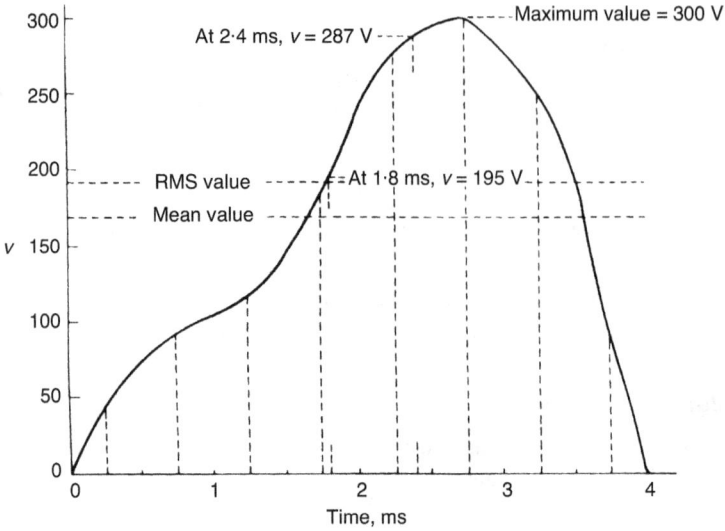

Figure 12.2   Graph for Examples 12.2–12.4

maximum value and the mean value of voltage.

$$f = \frac{1}{T} = \frac{1}{8} = \frac{1}{0.008} = 125 \text{ Hz}$$

The next step in the solution is to draw the half-cycle as a graph (Figure 12.2), reading off the instantaneous values (195 V at 1.8 ms, 287 V at 2.4 ms) and its maximum value (300 V).

To find the average or mean value, the baseline (time axis) is divided into any number of equal parts. For clarity, eight parts have been chosen, although more would give greater accuracy. At the centre of each part, a broken line has been drawn up to the curve.

The average value of voltage will be the average length of these lines (expressed in volts). To find this, we add the voltage represented by each line and divide by the number of lines.

$$U_{av} = \frac{45 + 91 + 118 + 185 + 278 + 300 + 248 + 85}{8} = \frac{1350}{8} = 169 \text{ V}$$

Effective value

Since the heat dissipated by a current is proportional to its square ($P = I^2 R$), the average value of an alternating current is not the same as the DC that produces the same heat or does the same work in the same time. The equivalent to a DC is the value we use most in describing and calculating AC systems and is called the effective or root mean square (RMS) value of the system. The RMS value is the square root of the mean value of the squares of the instantaneous values. In other words:

1. Find the instantaneous values.
2. Square each one.
3. Calculate the mean value of these measurements.

The symbols used for RMS values are U, I and so on. The method of finding the effective or RMS value of a given waveform is illustrated in the following example.

Example 12.3
Find the RMS value of the voltage waveform of Example 12.2
    To find the RMS value,

(a) Divide the base into equal parts and erect a vertical line to the curve from the centre of each part (as for finding the average value).
(b) Square the value of each vertical line.
(c) Take the mean of the squared values (add them and divide by the number of lines).
(d) Take the square root of the result The answer is the root of the mean of the squared value – Root Mean Squared.

The graph has already been drawn and vertical lines have been erected for Example 12.2, and need not be repeated in this case.
    The sum of the squared values in Example 12.2 will be:

$$45^2 + 91^2 + 118^2 + 185^2 + 278^2 + 300^2 + 248^2 + 85^2 = 294,468 \text{ V}^2$$

$$\text{Mean of the square values} = \frac{294,468}{8} = 36,809 \text{ V}^2$$

$$\text{Root mean square value} = \sqrt{36,809}\text{V} = 191.9 \text{ V}$$

It will be seen that the RMS value is greater than the mean value, and this is always the case, except for a DC, for which they are equal.

Form factor

Form factor is an indication of the shape of a waveform; the higher its value the more 'peaky' the waveshape. Form factor for a particular waveform is the ratio of the RMS and mean values:

$$\text{Form factor} = \frac{\text{RMS value}}{\text{Mean value}}$$

Example 12.4
Find the form factor of the waveform of Example 12.2.

$$\text{Form factor} = \frac{\text{RMS value}}{\text{Mean value}} = \frac{191.9}{169} = 1.136$$

12.4   Sinusoidal waveforms

In Chapter 9, we considered a simple rectangular loop of wire rotating on an axis between the poles of a permanent magnet. The EMF induced in the loop is shown again in Figure 12.3, with one cycle being induced for each revolution of the loop. If the loop rotates at a constant speed, the horizontal axis can be divided into equally spaced units of time as well as degrees of rotation.

For the production of DC, the generator is fitted with a segmented cylinder called a commutator.

- The ends of the armature loop are connected to a segment each and insulated from one another.
- The generated electricity is picked up by carbon brushes that press against the commutator.
- Each brush will only pick up the current generated as first one, then the other half of the loop will pass through the generator's magnetic field in a certain direction.
- This means that each brush will either be positive or negative.

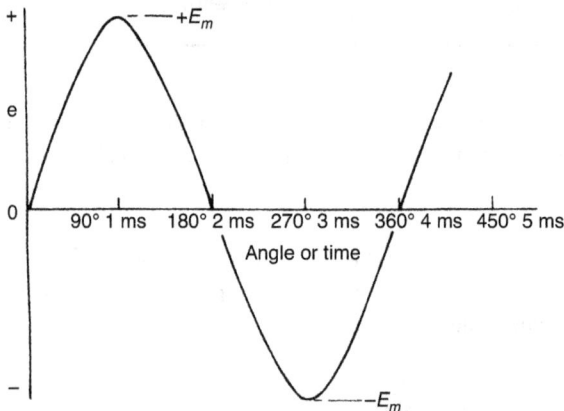

Figure 12.3   Sinusoidal waveform

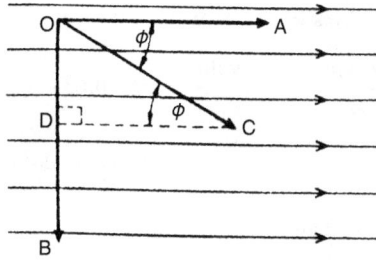

Figure 12.4   Effect of conductor direction on induced EMF

An AC generator is fitted with slip rings, which, in effect, are a separate commutator for each end of the armature loop.

- Each slip ring remains connected to its end of the loop for the entire cycle.
- As the loop passes through the field in one direction, it may be positive then, as it completes its cycle by passing through the field in the opposite direction, it will become negative.
- As a result, each half of the loop will produce current of alternating polarity e.g. + - + -.

The EMF induced in the loop at any instant depends on the rate of cutting lines of magnetic flux. Referring to Figure 12.4:

- Movement from O to A induces no EMF, whereas movement from O to B induces a maximum EMF, which we will call E_m.
- Moving the same distance from O to C at an angle of φ degrees to OA will induce an EMF proportional to the length OD.
- OD and OC are the opposite side to φ and the hypotenuse, respectively, of the right-angled triangle OCD.
- From simple trigonometry, EMF induced in moving from O to C

$$= E_m \times \frac{OD}{OC} = E_m \sin \varphi$$

The waveshape of Figure 12.3, translated as a graph of Em sin φ, is referred to as a 'sine wave', and is said to be 'sinusoidal' in shape. This waveshape is easily expressed as a mathematical formula, and is similar to that obtained from practical generators. From now on we will consider all AC electrical systems to have sinusoidal wave shapes. *Note*: It is common to use the radian as a measure of angle (see Chapter 1).

The total angular movement after t seconds of a wire loop rotating at f revolutions per second and giving an output of f cycles per second will be $2\pi ft$ radians. Thus,

$$e = E_m \sin 2\pi ft$$

$$e = E_m \sin \omega t$$

where ω (Greek letter 'omega') $= 2\pi f$ radians per second.

The average and RMS values of a sine wave are of importance. They are

$$\text{Mean value} = \frac{2 \times \text{Maximum value}}{\pi} \quad \text{or} \quad 0.637 \times \text{Maximum value}$$

$$\text{RMS value} = \frac{\text{Maximum value}}{\sqrt{2}} \quad \text{or} \quad 0.707 \times \text{Maximum value}$$

$$\text{Form factor} = \frac{0.707 \text{ Maximum}}{0.637 \text{ Maximum}} = 1.11$$

Value for alternating systems are always given as RMS unless otherwise stated.

Example 12.5
Find the maximum and mean values for a 230 V supply.

$$\text{RMS value} = 0.707 \times \text{maximum value}; \quad \text{therefore, } U_m = \frac{\text{RMS}}{0.707} = \frac{230}{0.707}$$

$$= 325 \text{ V}$$

$$\text{Mean} = 0.637 \times U_m = 0.637 \times 325 = 207 \text{ V}$$

$$\text{or Mean} = \frac{\text{RMS}}{\text{form factor}} = \frac{230}{1.11} = 207 \text{ V}$$

12.5   Phasor representation and phase difference

Wave diagrams, examples of which are shown in Figures 12.1–12.3, are an extremely useful way of depicting alternating values, but they are tedious to draw to scale. An alternative method of representing an alternating quantity is a straight line called a phasor. Its length is proportional to the value represented.

The phasor is assumed to pivot at the end without an arrowhead, and to revolve anticlockwise once for every cycle of the system it represents (Figure 12.5). If the vertical height of its moving tip at various instants is transferred to a graph as shown, a sinewave results.

Now consider two alternating quantities (Figure 12.6(a)).

- Quantity X passes through zero, going positive 45° after quantity Y, so we say that X **lags** Y by 45°.
- Alternatively, 'Y leads X by 45°. The angle of 45° between the two quantities is the phase angle, and if it is unknown, it is denoted by the symbol φ (small Greek letter 'phi').
- The phasor diagram for the arrangement is shown in Figure 12.6(b).

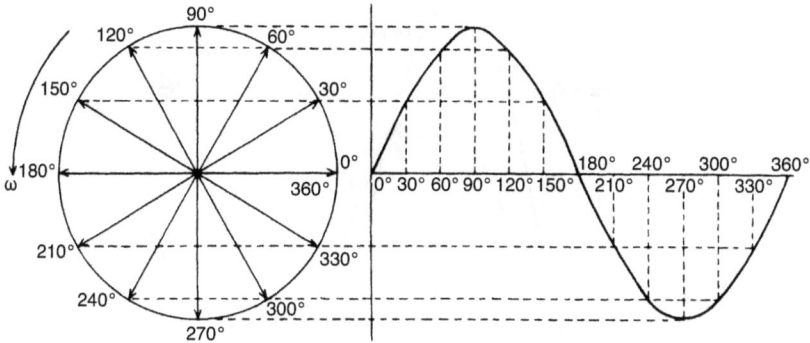

Figure 12.5   Representation of sine wave by rotating phasor

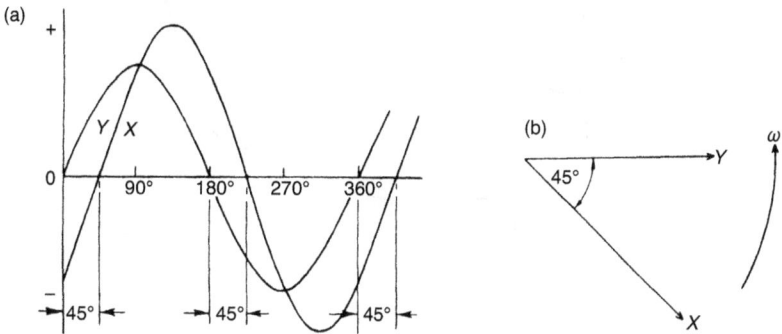

Figure 12.6   Two sine waves: (a) wave diagrams of two alternating quantities, X
lagging Y by 45° and (b) phasor diagram for waves of (a)

• The lengths of the phasors represent the RMS values of the alternating
  quantities.
• The angle between is the same as the phase angle of 45°, with Y leading X, and
  both rotating anticlockwise.

It is often necessary to add together two alternating values. If they are in phase (that
is, if there is no displacement of phase between them), they can be added in the
same way as DC values. For example, voltages of 100 and 150 that are in phase will
sum to 250 V.

   If the two values are not in phase, they can be added using a wave or a phasor
diagram, but not by simple arithmetic. Figure 12.7(a) gives the wave diagram of a
voltage U_1 of maximum value 100 V and U_2 voltage of maximum value 150 V,
which lags U_1 by 60°.

   The sum of these two waves at any instant is found by adding together the
instantaneous values of the individual waves for that instant. For example,

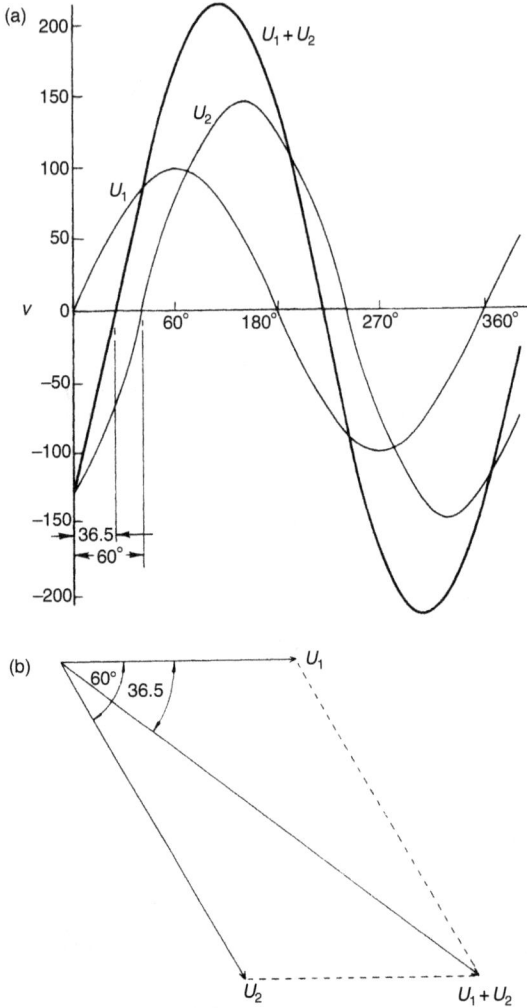

Figure 12.7   *Addition of alternating values by (a) wave diagram and (b) phasor diagram*

at $0°$, $v_1 = 0$

at $30°$, $v_1 = 50$

$v_2 = -130$ V so $v_1 + v_2 = -130$ V

$v_2 = -75$ V so $v_1 + v_2 = -25$ V

at $60°$, $v_1 = 86.6$ $v_2 = 0$ V

so $v_1 + v_2 = +86.6$ V

These two values can be added more quickly by using a phasor diagram (Figure 12.7(b)). U_1 and U_2 are first drawn to scale and their phasor sum $U_1 + U_2$ is found by completing the parallelogram as indicated (see Section 3.5 for instructions on completing the parallelogram). The line whose length and direction indicate an alternating current is the phasor (for many years referred to as a vector, and the terms 'vector' and 'vector diagram' are still sometimes used).

12.6   Resistive AC circuit

If an alternating voltage of $v = U_m \sin\omega t$ is applied to a resistor, the instantaneous current is:

$$i = \frac{v}{R} = \frac{U_m \sin\omega t}{R}$$

The current will be given by

$$i = I_m \sin\omega t, \text{ so } I_m \sin\omega t = \frac{U_m}{R}$$

or, by using RMS values, $I = \dfrac{U}{R}$

There is no phase difference between v and i, and the circuit calculations will be carried out in the same way as for a DC circuit. Circuit, wave and phasor diagrams are shown in Figure 12.8.

Example 12.6
A 230 V AC supply is connected to an 80 Ω resistor. Calculate the resulting current flow.

$$I = \frac{U}{R} = \frac{230}{80} = 2.88 \text{ A}$$

The current and voltage used are RMS values

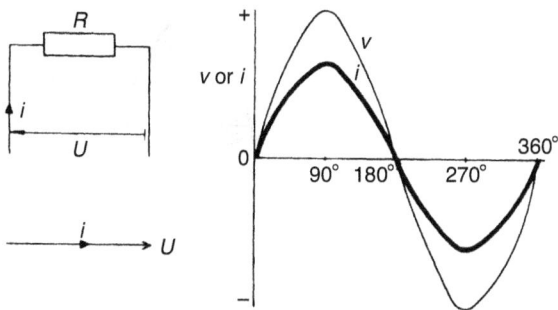

Figure 12.8   Circuit, wave and phasor diagrams for resistive AC circuit

12.7   Inductive AC circuit

The property of self-inductance was considered briefly in Chapter 9 and will be more fully discussed in Volume 2. Briefly, any coil of wire that sets up a magnetic field when it carries a current has this property, so that motor windings, relay coils and transformers possess self-inductance (symbol L).

If current changes in a winding, a second EMF will be induced to oppose the change. If the current is increasing, the EMF will oppose the supply voltage to limit the rate of increase, and if decreasing will try to keep the current flowing. The unit of self-inductance is the henry (symbol H).

Every coil of wire must possess resistance, and because of this resistance it is not practicable to produce a non-resistive or 'pure' inductance. However, in this section, we shall assume that a pure inductance exists and examine the result of applying an alternating voltage to it.

Circuit, wave and phasor diagrams are shown in Figure 12.9. The EMF induced in the winding must be in opposition to the applied voltage, so on the wave diagram v and e are drawn with a phase displacement of 180°. The induced EMF depends on the rate of change of current:

$$e = \frac{L(I_1 - I_2)}{t}$$

When e is zero, the rate of change of current must also be zero, a sinusoidal varying value, this only occurs at the maximum points, so current must be at maximum when e is zero.

Induced EMF must also be maximum when the rate of change of *current* is maximum. Since this occurs as the current passes through zero, maximum EMF must coincide with zero current.

When the current is going positive, the EMF induced must oppose this change of current and will therefore be negative. The current wave diagram can therefore be drawn and can be seen to lag the applied voltage by 90°. The phasor diagram can therefore be drawn as shown, with induced EMF omitted.

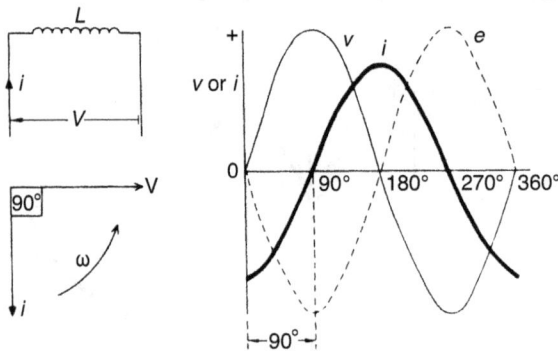

Figure 12.9   Circuit, wave and phasor diagrams for purely inductive AC circuit

We have assumed that the inductive circuit has no resistance, but since the resulting current flow is not infinite, it must be limited by some property other than resistance. This property is called the inductive reactance of the coil (symbol X_L), and it can be shown that:

$$X_L = \frac{U}{I} = 2\pi f L = \omega L$$

where X_L = inductive reactance of the coil; U = voltage applied to the coil, V; I = resulting current flow, A; f = supply frequency, Hz; L = coil inductance, H; $\omega = 2\pi f$.

Note that when $f = 0$, the inductive reactance will be zero. This means that the inductance of a coil has no effect on the steady flow of DC through it. That is limited only by the coil resistance.

Example 12.7
A coil has a self-inductance of 0.318 H and negligible resistance. Calculate its inductive reactance and the resulting current if connected to

(a) a 230 V 50 Hz supply;
(b) a 100 V, 400 Hz supply.

(a) $X_L = 2\pi f L = 2\pi \times 50 \times 0.318 = 100 \ \Omega$

$$I = \frac{U}{X_L} = \frac{230}{100} = 2.3 \text{ A}$$

(b) $X_L = 2\pi f L = 2\pi \times 400 \times 0.318 = 800 \ \Omega$

$$I = \frac{U}{X_L} = \frac{100}{800} = 0.125 \text{ A}$$

12.8 Capacitive AC circuit

Capacitance will be treated in greater detail at the beginning of Volume 2. A capacitor is a device that is capable of storing an electric charge when a potential difference is applied to it, and consists of two conducting plates that are very close together, but are separated by an insulator called the dielectric.

The symbol for capacitance is the letter C. The unit of capacitance is the farad (symbol F), but this unit is too large for most practical purposes, so the millionth part of a farad, or microfarad (symbol μF) is usual.

$1 \ \mu F = 1 \times 10^{-6}$ farad

Smaller units of capacitance are also used, notably the nanofarad (nF) and the picofarad (pF)

$1nF = 1 \times 10^{-9}$ farad

$1pF = 1 \times 10^{-12}$ farad

If a direct voltage is connected to a capacitor, negatively charged electrons will be attracted to the plate connected to the positive terminal of the supply and will be repelled from the negative terminal onto the other plate. Thus, one plate will have a surplus of electrons, and the other a deficiency of electrons.

In this condition, the plate system is said to be charged, and the potential difference between the plates will increase until it equals the supply voltage. At that point, the drift of electrons will cease.

• If a capacitor is connected to a DC supply, the flow of current will die away quickly as the capacitor charges.
• If supply is from an AC source, the capacitor will alternately charge and discharge with opposite polarity. Thus, although no current actually passes through the capacitor, a measurable alternating current exists in the circuit.

Capacitor charging

When an alternating voltage is applied to an uncharged capacitor and passes through zero going positive, the current will immediately reach its maximum value as the capacitor starts to charge.

As the charge increases, charging current will fall. It will reach zero when the voltage becomes steady, which will be for an instant at its maximum value.

As the voltage falls, the capacitor will discharge and a negative current results. This pattern repeats as demonstrated in the waveform (Figure 12.10), which shows us that in a capacitive circuit, current leads supply voltage by 90° (compare with Figure 12.9, which shows current lagging supply voltage by 90° in a purely inductive circuit).

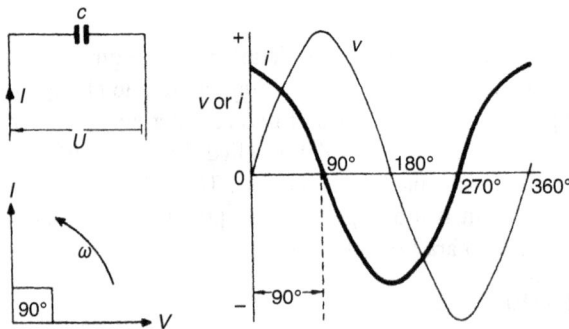

Figure 12.10   Circuit, wave and phasor diagrams for capacitive AC circuit

Calculating capacitive reactance

The flow of alternating current to a capacitor is determined by its capacitive reactance (symbol X_c).

$$X_c = \frac{U}{I} = \frac{1}{2\pi fC} = \frac{1}{\omega C}$$

where X_c = capacitive reactance; U = supply voltage, V; f = supply frequency, Hz; I = circuit current, A; C = circuit capacitance, F.

$$\omega = 2\pi f$$

Since the capacitance of capacitors is more often measured in microfarads, the expression can be written as

$$X_c = \frac{10^6}{2\pi fC}$$

where C = circuit capacitance, µF.

Example 12.8

Calculate the capacitive reactance at 50Hz of the following capacitors:

(a) 2 µF
(b) 8 µF
(c) 25 nF

Calculate also the current that will flow to each when connected to a 230 V, 50 Hz supply.

(a) $X_c = \dfrac{10^6}{2\pi fC} = \dfrac{10^6}{2\pi \times 50 \times 2} = 1590 \ \Omega$

$I = \dfrac{U}{X_c} = \dfrac{230}{1590} = 0.145 \ \text{A}$

(b) $X_c = \dfrac{10^6}{2\pi fC} = \dfrac{10^6}{2\pi \times 50 \times 10} = 398 \ \Omega$

$I = \dfrac{U}{X_c} = \dfrac{230}{398} = 0.578 \ \text{A}$

(c) $X_c = \dfrac{10^9}{2\pi fC} = \dfrac{10^9}{2\pi \times 50 \times 25} = 127 \ \text{k}\Omega$

$I = \dfrac{U}{X_c} = \dfrac{230}{127,000} = 1.81 \ \text{mA}$

12.9   Impedance

Impedance is looked at in more detail in Volume 2, but because it has a bearing on aspects of electrical supply and shock protection, it is necessary to introduce the value here. For a resistor, the effect limiting current for a given voltage is resistance

$$R = \frac{V}{I}$$

For a pure inductor, the effect is inductive reactance,

$$X_L = \frac{V}{I}$$

It can be seen that the ratio V/I is the effect limiting the current in a circuit. For a circuit including resistance and inductive reactance this effect is due to a combination of both.

For an AC circuit, this combination is the sum of these two values and is called impedance (symbol Z) that is measured in ohms.

$$Z = \frac{V}{I}$$

$$Z^2 = R^2 + X^2 \text{ or } Z = \sqrt{(R^2 + X^2)}$$

Example 12.9
A 50 Hz, 230 V inductive circuit has a resistance of 125 Ω, and an inductance of 1.3 H. Calculate

(a) Impedance
(b) Current

(a) Step 1 – calculate X_L

$$X_L = 2\pi fL \ X_L = 2\pi \ x \ 50 \ x \ 1.3 = 94.26 \ \Omega$$

Step 2 – Calculate Z

$$Z = \sqrt{R^2 + X^2} \ Z = \sqrt{125^2 + 94.26^2} = 156.56 \ \Omega$$

(b) $I = \dfrac{V}{Z} \ I = \dfrac{230}{156.56} = 1.47 \ A$

12.10   Summary of formulae for Chapter 12

For sinusoidal waveforms

$$\text{Periodic time } T = \frac{1}{f} \quad f = \frac{1}{T}$$

where T = periodic time, s, and f = frequency, Hz.

$$\text{Mean value} = \frac{2 \times \text{Maximum value}}{\pi} \text{ or } 0.637 \times \text{Maximum value}$$

$$\text{RMS value} = \frac{\text{Maximum value}}{\sqrt{2}} \text{ or } 0.707 \times \text{Maximum value}$$

$$\text{Form factor for any waveshape} = \frac{\text{Effective value}}{\text{Mean value}}$$

For purely resistive AC circuits

$$I = \frac{U}{R} \quad R = \frac{U}{I} \quad U = IR$$

where U = applied voltage (effective value), V; I = resulting current (effective value), A; and R = circuit resistance.

For a purely inductive AC circuit

$$I = \frac{U}{X_L} X_L = \frac{U}{I} \quad U = IX_L$$

where X_L = inductive reactance, Ω.

$$X_L = 2\pi f L \quad f = \frac{X_L}{2\pi L} \quad L = \frac{X_L}{2\pi f}$$

where f = supply frequency, Hz, and L = circuit inductance, H.

For a purely capacitive AC circuit

$$I = \frac{U}{X_c} \quad X_c = \frac{U}{I} \quad U = IX_c$$

where X_c = capacitive reactance, Ω

$$X_c = \frac{1}{2\pi f C} \quad f = \frac{1}{2\pi C X_c} \quad C = \frac{1}{2\pi f X_c} \quad X_c = \frac{10^6}{2\pi f C}$$

where circuit capacitance is in μF.

Impedance

$$Z = \sqrt{R^2 + X^2} \text{ or } Z^2 = R^2 + X^2 \quad Z = \frac{V}{I} \quad I = \frac{V}{Z} \quad V = IZ$$

where Z = impedance, Ω

$X = Xc,\ X_L$ or $X_L - Xc,$ Ω

12.11   Exercises

1. An alternating voltage completes one cycle in 1 ms. What is its frequency?
2. Explain the meaning of the term 'periodic time' as applied to an AC system. What is the periodic time of a system with a frequency of 60 Hz?
3. An alternating voltage is triangular in shape, rising at a constant rate to a maximum of 300 V in 0.01 s, and then falling to zero at a constant rate in 0.005 s. The negative half-cycle is identical in shape to the positive half-cycle.
   Calculate
   (a) the supply frequency
   (b) the average voltage
   (c) the RMS voltage
   (d) the form factor.

4. A voltage is represented by a sine wave, and has a maximum of 200 V. Calculate its RMS and average or mean value.
5. Explain, with the aid of a diagram, the meaning of the following terms as applied to an alternating EMF:
   (a) maximum value
   (b) root means quare value
   (c) cycle.

   If the frequency of the supply is 50 Hz, what is the time taken for the completion of one cycle?
6. The positive half-cycle of a sinusoidal alternating current has the following values at 1 ms intervals:

T (ms)	0	1	2	3	4	5	6	7	8	9	10
I (A)	0	7	13	19	24.5	30.5	35.5	40	44.5	48	50.5
T (ms)	11	12	13	14	15	16	17	18	19	20	
I (A)	52	52.5	52.5	51.5	49.5	45.5	39.5	31	15	0	

Plot this curve, and hence calculate
(a) the supply frequency
(b) the mean value of the current
(c) the effective value of the current
(d) the form factor
(e) the instantaneous value of current after 27 ms.

7. Using sine tables, draw out the positive half-cycle of a sinusoidal voltage of peak value 200 V, and calculate the average voltage, the RMS voltage and the form factor.

8. A sinusoidal current has an effective value of 10 A. Calculate its average and maximum values.

9. Calculate the mean and peak voltages of 230 V sinusoidal supply.

10. Explain the following terms applied to an alternating-current wave:
   (a) maximum value
   (b) average value
   (c) RMS value.

   If the maximum value of a sine wave is 300 A, give the average and RMS values.

11. (a) Draw freehand two complete cycles of a sinusoidal alternating current of peak value 100 A. Label
   (i) one point of current zero
   (ii) one complete
   (iii) one negative peak.
   (b) What is the RMS value of this sine wave
   (c) Why is the RMS value (effective value) normally used to specify the current value?

12. Draw free hand two complete cycles of a sinusoidal alternating current. Label one negative peak, one positive peak, one current zero and one positive half-cycle. Why is an alternating current usually specified by its RMS, or effective, value? If the peak value of a sine wave of current were 200 A, what would be its RMS value?

13. Two sinusoidal currents, each of 100 A peak value, are added. One starts at zero time, and the other a quarter cycle later. Draw the two waves, and the wave of their sum, and measure
   (a) the peak value of the sum
   (b) the degrees after zero time at which the sum reaches its first peak. The following values may be used:

θ	0°	30°	60°	90°	120°	150°	180°
$\sin \theta$	0	0.5	0.866	1	0.866	0.5	0

14. Draw a scale phasor diagram to show a current of 15 A leading current of 10 A by 30°. Find the resultant of these phasors, and its phase relative to the 10 A current.

15. Find the resultant of two 100 V AC supplies which are connected in series but 120° out of phase
   (a) by phasor addition
   (b) by wave addition.
   Find the phase of the resultant relative to either 100 V supply.

16. Calculate the current in a 20 Ω resistor connected to a supply at
    (a) 230 V, 50Hz;
    (b) 115 V, 400 Hz
    (c) 1000 V, 100 Hz
    (d) 200 V, 30 Hz
    (e) 300 V, 60 Hz

17. Calculate the inductive reactance of the given non-resistive inductor, and the current connected to the given supply:
    (a) 1 H connected to a 230 V, 50 Hz supply
    (b) 20 mH connected to a 400 V, 60 Hz supply
    (c) 0.15 H connected to a 100 V, 400 Hz supply.

18. Calculate the capacitive reactance of the given capacitor and the current when connected to the given supply.
    (a) 10 μF connected to a 230 V, 50 Hz supply
    (b) 20 nF connected to a 12 V, 1 kHz supply
    (c) 300 μF connected to a 3V, 30 Hz supply
    (d) 1 pF connected to a 24 V, 50 kHz supply
    (e) 40 nF connected to a 1 V, 1 MHz supply

19. Calculate the impedance of an inductive AC circuit if the resistance is 83 Ω, the frequency is 50 Hz and the inductance is 0.3 H

20. Calculate the impedance of an inductive 60 Hz AC circuit if the resistance is 42 Ω, X_L is 85 Ω and X_c is 28 Ω

12.12 Multiple-choice questions

12M1 An alternating current is best defined by a graph that will indicate the
    (a) positive half-cycles only
    (b) negative half-cycles only
    (c) waveform
    (d) RMS value

12M2 The number of cycles of an alternating system traced out each second is called the
    (a) periodic time
    (b) waveform
    (c) half-cycle time
    (d) frequency

12M3 The periodic time of an alternating current is the
    (a) time taken for completion of one cycle
    (b) time before a positive half-cycle falls to zero
    (c) time taken for completion of a half-cycle
    (d) usually about one second

12M4   The frequency of an alternating electrical system is measured in
(a) cycles
(b) volts
(c) hertz
(d) seconds

12M5   If the periodic time of an alternating-current system is 120 μs the
frequency is
(a) 8 kHz
(b) 0.000125 s
(c) 0.008 c/s
(d) 8 Hz

12M6   A major advantage of an alternating-current supply over a direct-current
system is that
(a) electric shocks are likely to be less severe
(b) the cost will be lower
(c) the current has many values instead of only one
(d) transformers will not operate efficiently on direct-current supplies

12M7   The highest value reached by an alternating voltage in a half-cycle is called
the
(a) maximum or peak value
(b) root mean square value
(c) instantaneous value
(d) average value

12M8   The average value of an alternating system is taken for one half-cycle only
because
(a) the calculation is impossible for a full cycle
(b) otherwise positive and negative half-cycle values will cancel
(c) there is a shock danger if the full cycle is considered
(d) if a full cycle is considered the result is the effective value

12M9   The RMS (root mean square) value of an alternating current is
(a) the peak value of a square-wave system
(b) seldom used because it is so difficult to calculate
(c) the effective value that is almost always given
(d) the square root of the average value

12M10 The form factor of an alternating waveform is
(a) found by dividing effective value by average value
(b) an indication of the 'peakiness' of the waveshape
(c) is the ratio of maximum and RMS values
(d) is the ratio of maximum and average values

12M11 An alternating-voltage waveform can be expressed by the formula $e = E_m$ sin$2\pi ft$ and is called
(a) a trigonometrical waveform
(b) a phasor
(c) a sinusoidal voltage waveform
(d) a radian

12M12 A phasor is
(a) part of a three-phase system
(b) a method of representing a sinusoidally varying system
(c) the angle between alternating current and voltage
(d) always measured in radians

12M13 The angle between two alternating quantities is called
(a) the lag angle, represented by π
(b) the phase angle, and is fixed at 30°
(c) the angle of lead, represented by μ
(d) the phase angle, represented by φ

12M14 Two phasors can be added by
(a) completing the parallelogram
(b) adding their lengths with a calculator
(c) using a wave diagram
(d) drawing them end to end and measuring the result

12M15 The RMS value of a sinusoidal system can be found
(a) by multiplying its maximum value by π
(b) by dividing the average value by the form factor
(c) by multiplying the average value by the form factor
(d) by dividing the maximum value by $\sqrt{2}$

12M16 If a sinusoidal voltage is applied to a pure inductor
(a) the average current flow will be zero
(b) the current will lead the voltage by 90°
(c) the current will lag the voltage by 90°
(d) current and voltage will be in phase

12M17 The effect that limits current flow in a pure inductor connected to an alternating-voltage supply is called the
(a) induced EMF
(b) inductive reactance
(c) capacitive reactance
(d) resistance

12M18 If a supply at 230V, 50Hz is connected to a 420 mH conductor, the current flow will be
    (a) 1.74 A
    (b) 0.57 A
    (c) 11.4 A
    (d) 31.6 µA

12M19 If a sinusoidal voltage is applied to a pure capacitor
    (a) the average current flow will be zero
    (b) the current will lead the voltage by 90°
    (c) the current will lag the voltage by 90°
    (d) current and voltage will be in phase

12M20 An alternating supply of 12 V at 1 kHz is connected to a capacitor, when the current is found to be 15 mA. The capacitance of the capacitor is
    (a) 800
    (b) 625 nF
    (c) 50.3 µF
    (d) 198 nF

Chapter 13

Supply and earthing

13.1   Introduction

The purpose of the next three chapters is to show how the basic theory we have already considered is put into practice to provide electrical supplies and systems.

13.2   Generation and distribution

The main source of electricity are power stations. The major power stations are distributed over the country and feed into a nationwide grid system which transmits and distributes power.

Power stations tend to be identified by the heat source used. Until recently this included coal, but the last coal-fired power station has been shut down. Other fossil fuels used for generation are gas and oil. Nuclear energy is also used to produce the necessary heat. This is considered to be a clean energy; however, the storage of radioactive fuel rods is a major part of the process and requires secure facilities and highly specialised processing (Figure 13.1(a)).

There are an increasing number of alternative power stations, including small plants that feed a local area and those that burn waste products as a heat source (Figure 13.1(b)). This is called biomass heating. In some cases, the steam from these systems is recycled to provide heat and hot water to the locality (Power Led – Combined Heat and Power [CHP]).

Figure 13.1   (a) Sizewell power station, image credit: N. Chadwick, via Geograph, CC BY-SA 2.0. (b) Scottish and Southern Slough power station, image credit: Chris Allen, via Geograph, CC BY-SA 2.0.

Wind turbine and photovoltaic generation also play a significant part in national power generation. In some countries, geothermal heat is an energy source for electrical generation. These systems will be looked at in more detail in Volume 2.

Fossil fuel, biomass and nuclear power stations work on a similar principle:

1. The energy produced is used to heat water into dry steam, which means that it is so hot it will not condense until it has served its purpose.
2. The steam drives a **turbine** which, in turn, drives a generator.
3. The average power station output is 25 kV, which is transformed up to either 275 kV or 400 kV for the initial stage of its transmission.
4. The steam is released into the air once it has passed through the turbine. Alternatively, it can be pumped into the locality as a heat source (Power Led – CHP).

KEY TERM: Turbine – A propellor like component that is rotated by a mechanical energy source then, in turn, drives a machine such as a generator.

The 275 kV and 400 kV networks are called the supergrid. The high voltage is required to counteract the volt drop effect and energy losses ($P = I^2R$) that will occur when current travels a long distance. The components of the transmission and distribution system are:

- Supergrid – 275 kV or 400 kV.
- Grid Supply Points – strategically placed substations where the voltage is transformed down to 133 kV for transmission to the next stage.
- Further substations where the voltage is transformed down to 33 kV or 11 kV.
- Local substation – where 11 kV or 33 kV supplies are transformed down to 400/230 V and supplied to local homes and businesses.
- HV users – factories and other large industrial and commercial premises may require a high-voltage supply such as 1000 V. This will be tapped off the grid via their own substation and distributed around the site as a ring main. Transformers will reduce the voltage to the required values.

The main method used to transmit and distribute electricity nationally is via overhead lines supported by steel pylons. The uninsulated cables are kept out of reach by the pylon height, and access is restricted by barbed-wire skirts placed about the pylon's legs. The cables are supported by vertical insulators that hang from the pylon arms (Figure 13.2). In the past, these insulators were ceramic or glass, but more recent versions are plastic.

Supplies can also be run underground, but this is considered more expensive and disruptive to carry out over long distances. Supplies to premises from local substations, however, are usually routed underground.

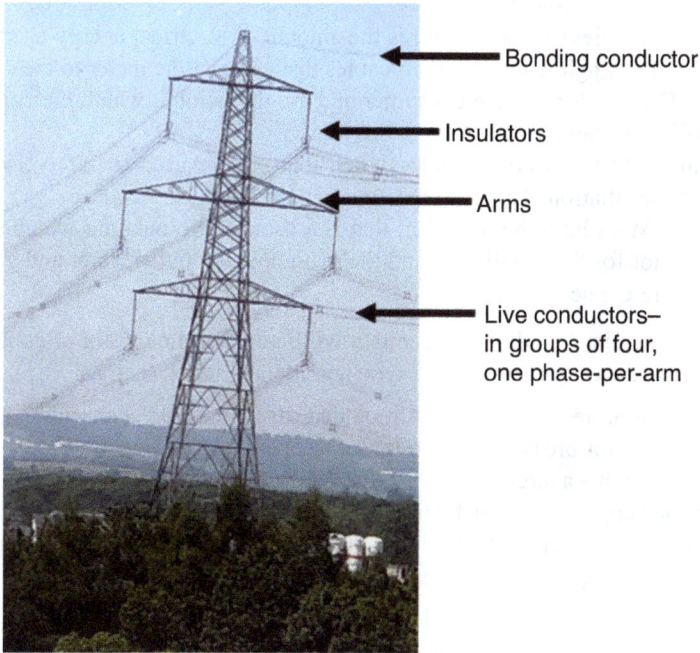

Bonding conductor

Insulators

Arms

Live conductors–
in groups of four,
one phase-per-arm

Figure 13.2   Labelled image of a pylon

The intake position

Once an electricity supply reaches a building, it is taken in via an intake position.
For a domestic property, this consists of the following (Figure 13.3):

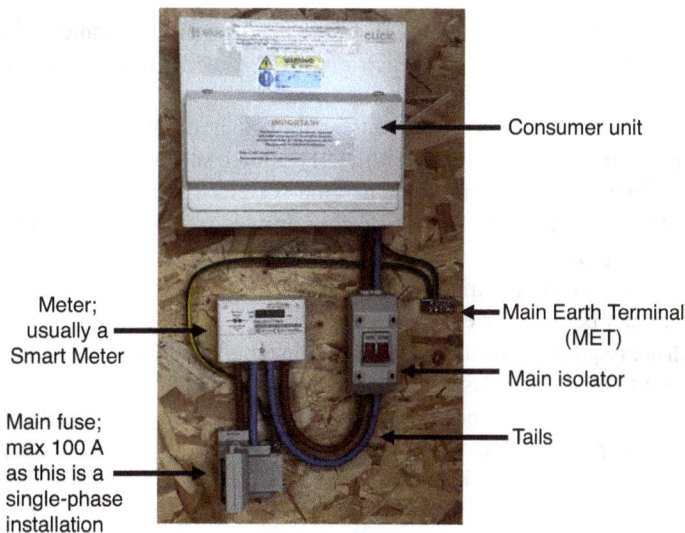

Consumer unit

Meter;
usually a
Smart Meter

Main Earth Terminal
(MET)

Main isolator

Main fuse;
max 100 A
as this is a
single-phase
installation

Tails

Figure 13.3   Typical domestic main intake position (by kind permission of
Oaklands College, St Albans)

- A main fuse – usually 100A
- Meter – the device that measures the amount of electrical energy used in kWh
- Tails – the single-core, sheathed cables that connect the meter to the consumer unit. These belong to the consumer and are the point at which the consumer's installation starts
- Main isolator – a double-pole switch that can be used to safely isolate the whole installation. This also belongs to the consumer
- MET – Main Earthing Terminal that is a connector block into which the earth conductor for the installation and the main bonding to the water and gas intake pipes are connected

A commercial or industrial premises will have the same chain of components, but they will differ in capacity and type and include items such as:

- The main protective device – a moulded case circuit breaker, which is a high-current circuit breaker that can be set to operate at the required fault current
- Fused switch – a large isolator switch that withdraws and inserts a rack of fuses into the supply. Used for high current loads
- Busbar – a method of supplying a number of distribution boards using copper bars and clamps rather than cables

Once these components would have been erected as separate items and wired together, but now it is common for them to be incorporated into a main panel. This panel will often provide other services such as voltage, current and power monitoring and environmental and supply optimisation controls. The panel may also incorporate the central processing unit for a building management system, used to monitor and control the power, environment, alarms and security in the premises. Building Management Systems are looked at in more detail in Chapter 15.

13.3   Direct-current supplies

Direct-current (DC) supplies are not available from supply companies. The reasons for this are listed in Chapter 12, which gives some of the more important advantages of AC systems.

Although mains supplies are all of the alternating-current (AC) type, it must be remembered that DC supplies are still in very wide use. For example, railway systems, ELV lighting, computers, fast charging for electric vehicles and emergency supplies where it is often most economical to provide a battery of secondary cells.

These cells provide a DC supply, usually for emergency lighting in the event of a mains failure. Again, the electrical systems of motor vehicles are almost always DC so that batteries can be used for engine starting and other services (radio, parking lights, etc.) which are required when the engine is not running. The modern practice is to fit an AC generator to motor vehicles, but the output is converted to a DC system before use.

Electric cars run on DC, but charging points are invariably fed by AC direct from the supply. This AC supply is then converted to DC via a rectifier (see Chapter 18).

Equipment intended to break current on DC supplies is always of heavier construction and more expensive than its AC counterpart. This is because the

Figure 13.4 DC isolator, or DC disconnector. Note grey and black colour scheme (by kind permission of IMO Technology).

voltage across a switch, or a similar break in a circuit, is continuous and tends to maintain any arc that may have formed as the contacts separated. Such an arc dissipates a great deal of heat and must be broken as soon as possible. Wide separation of switch contacts is required because of this. More complicated methods of breaking the arc, such as the provision of arc chutes or immersion of contacts in oil, are necessary in heavy-current systems.

DC isolators are generally called disconnectors (Figure 13.4).

13.4 Identification of fixed wiring by colour

The April 2004 amendments to the Wiring Regulations (BS 7671) unified the cable colours throughout Europe.

Function	Old colour	New colour
Single phase		
Phase	Red	Brown
Neutral	Black	Blue
Protective	Green/yellow	Green/yellow
Three phase		
Phase 1	Red	Brown or L1
Phase 2	Yellow	Black or L2
Phase 3	Blue	Grey or L3
Neutral	Black	Blue or N
Protective	Green/yellow	Green/yellow

New installations must use these colours, but there is no intention to rewire existing healthy circuits wired in pre-2004 colours. It follows that new extensions to existing three-phase installations will pose safety problems because the black conductor in the original installation will be a neutral whilst in a new extension it will be phase L2. To reduce this risk, a notice must be provided at the appropriate distribution boards to draw attention to the fact that two different systems of conductor identification exist within the installation.

13.5   Single-phase AC supplies

Most consumers are fed by means of a single-phase AC supply. Two wires are used:

- One called the line conductor and coloured brown (or red for older installations).
- The other is called the neutral conductor and coloured blue (or black for older installations).
- The earth wire is coloured green and yellow.

Both phase and neutral conductors are called 'live' because both normally carry current. The neutral is usually earthed at the supply source.

The majority of houses have a single-phase AC supply, which is fed in by means of a two-core cable when the supply is underground. The neutral or the cable armouring is used as the earth conductor, depending on the type of earthing system (described later in this chapter).

If the house is supplied via a pair of overhead conductors, earth will be provided by an installation earth electrode. The majority of single-phase supplies are obtained by connection to one phase of a three-phase system. The standard UK single-phase voltage is 230 V. The maximum installation current demand allowed for single-phase is 100 A. Any installation that exceeds this demand requires a three-phase supply.

13.6   Three-phase AC supplies

The standard method of generating, transmitting and distributing electrical energy in most countries is by use of a three-phase AC system.

To understand three-phase, it is necessary to see how it is generated. Remember that electromagnetic induction can take place if:

- A conductor is moved across lines of magnetic flux.
- A magnetic flux is passed over stationary conductors.
- Both magnet and conductor are stationary, but the flux strength is changed (e.g. transformer).

The three-phase generator uses the second of those methods (Figures 13.5 and 13.6):

- The stator (stationary) is wound with sets of three windings. These triplets consist of a winding for each phase. These are at an angle of 120° from each other.

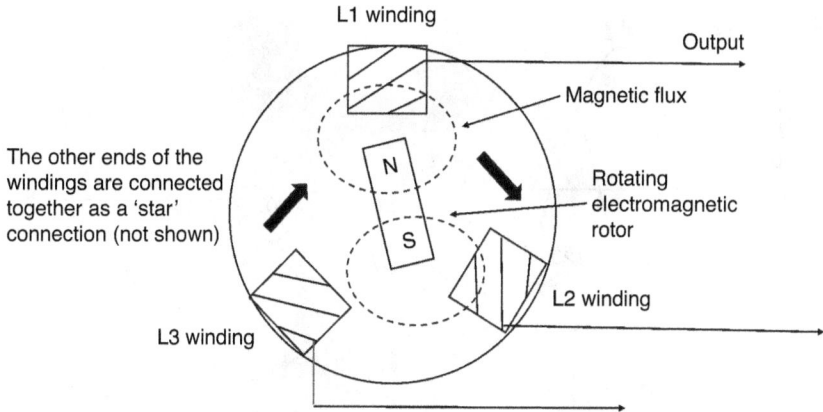

Figure 13.5   Simplified diagram of a three-phase generator to demonstrate operating principles. In reality, the windings are installed in pairs (pole pairs) and there may be more than a single pair for each phase.

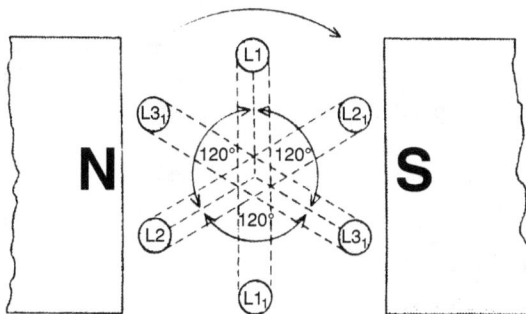

Figure 13.6   Arrangement of simple wire-loop generator to produce three-phase supply

- One end of each winding is connected to the generator output. The other is connected to the other two windings.
- The rotor, which is the rotating part inserted into the centre of the stator, takes the form of an electromagnet. A DC supply provides the excitation required to energise its magnetic field.
- As the rotor turns within the stator, its magnetic flux sweeps over the field windings and induces current into each winding in turn.

The phasor diagram for the induced EMFs is also shown (Figure 13.7). It will be seen that the electrical phase displacement between the EMFs is 120°, the same as the physical displacement of the field windings.

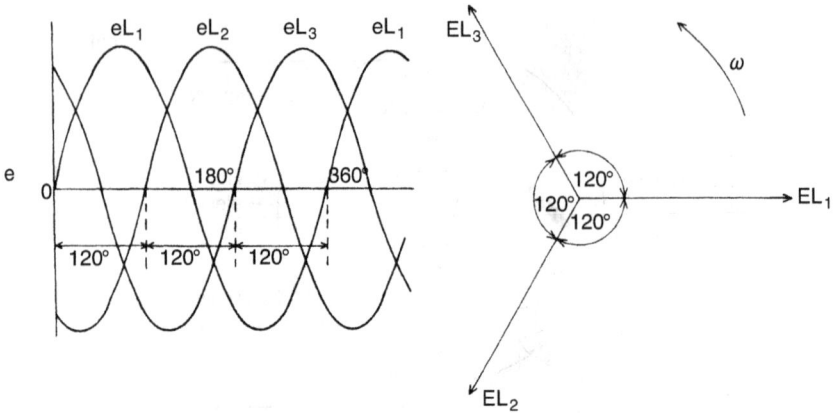

Figure 13.7   Wave and phasor diagrams for three-phase supply

There are a number of reasons for the adoption of three-phase transmission and distribution systems. Some of the most important are as follows:

• Load is shared over the three phases, which means smaller conductor sizes
• Three-phase motors have many advantages over single-phase motors, including smaller size, steady torque output, and the ability to self-start.
• When connected in parallel, single-phase generators present difficulties which do not occur with three-phase generators.
• Provides both three- and single-phase supplies.

The phases are called lines and are identified by the colours: L1 – brown; L2 – black; L3 – grey. Pre-2004, the colours were red, yellow and blue, respectively.

Three-phase windings (e.g. in transformers and the connections within some motor-starters) can be connected in two formats. See Figures 13.8 and 13.9.

• Delta – each phase is connected to the others, one at each end.
• Star – All the phases are connected, at one end only, to a central, or star, point. This point can provide a supply neutral terminal.

In practice, the windings are not laid out in neat star or triangle shapes as shown. However, the phases will be *connected* as shown.

The relationships between the phase and line voltages and currents are as follows.

STAR

$$U_L = \mathrm{Up}\sqrt{3} \quad \mathrm{Up} = \frac{U_L}{\sqrt{3}}$$

$$I_L = I_p$$

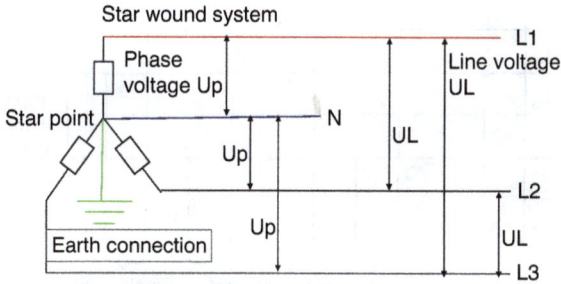

Figure 13.8 *Star configuration. Voltages between each phase and neutral are phase voltages (Up). Voltages between phases are line voltages (U_L). Note: neutral created at the star point.*

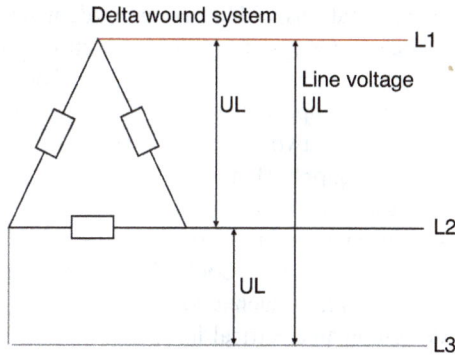

Figure 13.9 *Delta wound system. Note: absence of phase voltage because there is no facility for a neutral in this system.*

DELTA

$$U_L = U_p$$

$$I_L = I_p\sqrt{3} \quad I_p = \frac{I_L}{\sqrt{3}}$$

These relationships will be proved in Volume 2.

Three-phase supplies are taken into large industrial and commercial premises, and are used directly to supply heavy individual loads, such as large motors, heaters, furnaces and so on. If the load is unbalanced, that is if each of the three parts of the load takes a different current, a star connection with a neutral (known as a three-phase four-wire supply) is used. If the load is balanced, no neutral is needed and the load may be star- or delta-connected.

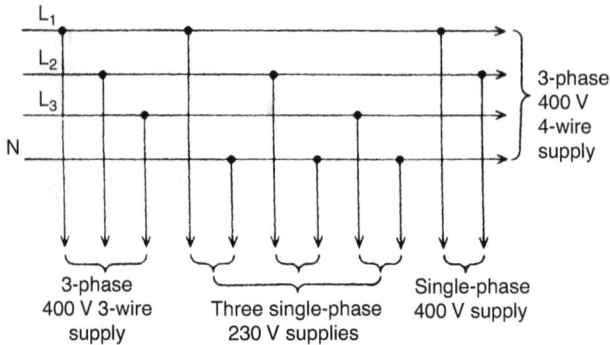

Figure 13.10   Supplies obtainable from standard three-phase four-wire supply

For most domestic applications with a maximum demand of under 100 A, the higher voltage of the three-phase system (400 V, which is $\sqrt{3} \times 230$ V, between lines) is not required, and single-phase supplies are taken from the three-phase four-wire supply. Each single-phase supply is connected to one phase and the neutral, so that multiples of three separate two-wire single-phase supplies can be obtained from a three-phase four-wire supply (Figure 13.10).

Even an industrial installation will require single-phase as well as three-phase. These 230 V supplies are used for lighting and general small-load purposes, such as cleaning, heating, ventilating and so on. Some single-phase welders operate at 400 V and the diagram also shows how such a connection can be made.

Care is taken in designing an electrical installation to 'balance the load'. To do this, the loads on the three separate phases are kept as nearly equal as possible, always bearing in mind that it may be dangerous for the areas served by different phases to overlap. This is because 400 V exists between single-phase circuits derived from different phases of a three-phase supply.

Example 13.1
The following single-phase circuits are required for a small engineering workshop on an industrial estate. The workshop has a three-phase supply so the 230 V circuits will be taken from each of the phases. Select the phases the circuits could be fed from to maintain a balance between phases. It is not always possible to achieve a perfect balance.

4 × ring final circuits –32 A each
3 × lighting circuits – 3.4 A, 2.8 A and 1.7 A, respectively
1 × water heater – 12 A
1 × small cooker – 15.5A
4 × single-phase machines – 9 A, 23 A, 18 A and 11 A, respectively

Suggested distribution of circuits:

L1	L2	L3
Ring final – 32 A	Ring final – 32 A	2 × Ring final – 64 A
Lighting – 3.4 A	Water heater – 12 A	Machine – 11.6 A
Lighting – 1.7 A	Machine – 23 A	
Cooker – 15.5 A	Machine – 9 A	
Machine – 18 A		
Lighting 2.8 A		
73.4 A	76 A	75.6 A

13.7 Supply earth

The final part of the electrical transmission and distribution system is the substation. From there a three-phase supply is taken out to the surrounding area to feed the various premises in need of electricity. The main item of equipment in the substation is a three-phase transformer. The primary side is wired in a delta configuration and the secondary in star.

Figure 13.8 shows that the star wound system has a star point at which all three phases meet. In a substation transformer, the star point is where the transformer is connected to earth. It is also the point at which the neutral is created. This means that current flows into an installation via the line conductor and then out through the neutral and down to earth.

At this point, earth acts as the negative terminal for electrical supplies. The general mass of earth is made up almost entirely of materials that are reasonable electrical conductors themselves or are made so because of its water content. From this, it follows that a current will flow to earth through a conductor which connects a live system to earth, provided that some other point of the system at a different potential is also connected to earth.

13.8 Shock protection

As we have seen, the earth from an installation is ultimately connected to the star point at the substation transformer. This provides the path along which fault currents can flow safely to earth. The purpose of earthing is to protect humans from electric shock.

Figure 13.11 shows a simplified circuit diagram for the supply from the substation to the installation. A fault exists between the line and the metal frame of an electrical appliance. Note that the earth conductor is connected to the metalwork of that equipment. The fault current flows down the earth conductor, back to the substation where it is connected to the earth itself. This forms a loop.

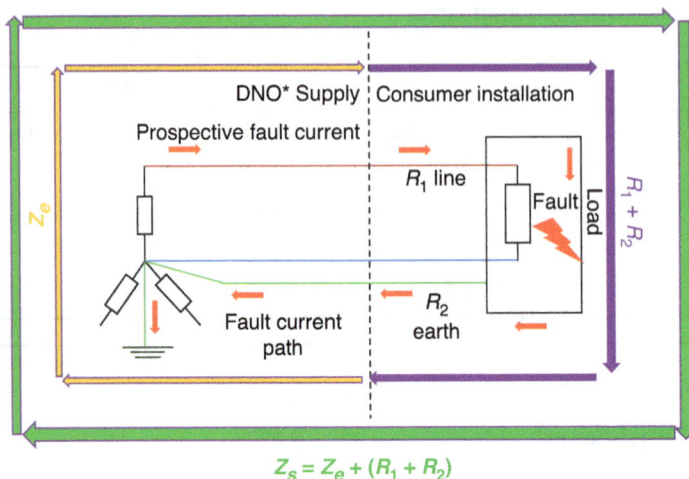

$$Z_s = Z_e + (R_1 + R_2)$$

Figure 13.11   The components of the earth fault loop path

- Fault current to star point and supply earth
- High fault current from transformer to installation and the point of fault
- High fault current operates protective device

If a human body forms part of this loop by touching the phase conductor while also in contact with earth, the person concerned will receive a shock. This circuit is shown in Figure 13.11. This possibility of shock is the disadvantage of earthing, but is comparatively easy to prevent by what BS 7671 defines as basic protection, that is, protection against shock under normal, no-fault conditions. Methods of basic protection:

- Insulation on conductors
- Enclosures – e.g. a socket or switch box with screw-on cover
- Barriers – e.g. bus bar cover in a consumer unit
- Obstacles – the barbed-wire skirts around the legs of a pylon
- Placing out of reach – uninsulated high-voltage cables supported by pylons

Removal of basic protection must be by a tool, key or by destruction (e.g. stripping cable insulation).

All exposed conductive metalwork of an installation must be connected to earth. BS 7671 defines exposed conductive parts as the metalwork directly associated with an electrical installation. For example:

- Steel conduit and trunking
- Metal switch and socket outlet boxes
- Metal consumer units
- Electric cookers, washing machines and tumble dryers
- Electric motors

The conductor that connects this metalwork to the earth terminal at the circuit source is called the circuit protective conductor (cpc). This can be:

- A separate conductor
- Armouring of a cable such as steel-wire armoured
- Steel conduit

13.9 Earthing systems

The consumer unit is, in turn, connected to the supply earth via the earth conductor. This will be connected to the supply earth by one of the following methods:

- The supply cable sheath (TN-S)
- The supply neutral (TN-C-S)
- An installation earth electrode (TT)

The TN-S system (see Figure 13.12) connects the installation to the substation star point via a separate conductor. This is usually the armouring of the supply cable.

The TN-C-S system (see Figure 13.13) connects the installation to the sub-station earth via the supply neutral. The connection is made at the intake position.

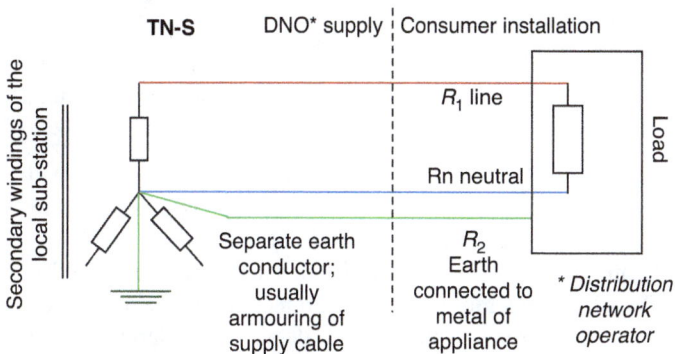

Figure 13.12   Terra neutral-separated (TN-S) earthing system

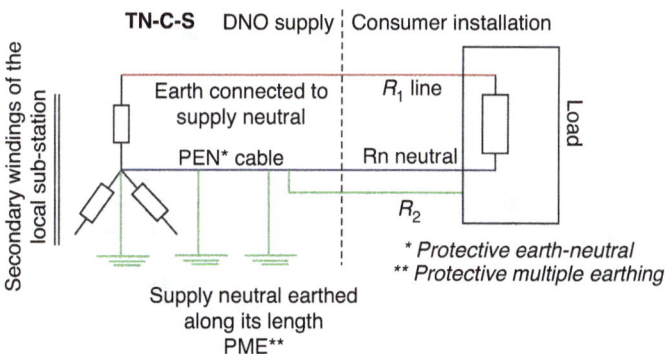

Figure 13.13   Terra neutral-combined-separated (TN-C-S) earthing system

The neutral is earthed along its length as a safety precaution. This is called Protective Multiple Earthing (PME). The reason for combining the earth and neutral is to turn an earth fault into a short circuit between the line and neutral. The fault current unleashed by a short circuit is generally higher than an earth fault. This means that the protective device will operate more quickly.

The terra-terra earthing system (see Figure 13.14) connects the installation to the mass of earth via an electrode at the installation itself. The electrode may vary from a complex grid of buried conducting plates or tapes to a simple rod driven into the ground. The IET On-Site Guide and Guidance Note 8 list the following as suitable earth electrodes. These include:

- Rod
- Pipes
- Tapes
- Plates
- Underground structural steelwork

By their very nature, earth electrodes are not always a reliable earth connection. Their effectiveness depends on the resistance between the metal of the rod and the material of the earth into which they have been driven. This will vary with the weather and temperature.

- Damp weather will lower the resistance as the soil's water content increases.
- Dry weather will raise the resistance as the soil may become crumbly and, as a result, a poorer conductor.

Earth electrodes must be tested. Testing is described in greater detail in Book 2. The maximum resistance at which an earth electrode is considered to be stable is 200 Ω between the electrode metal and the soil.

To mitigate against the varying resistance and effectiveness of an earth electrode, all TT installations must be further protected by a Fault Protection Residual Current Device (RCD; see Chapter 14 for a detailed description of RCDs and other protective devices).

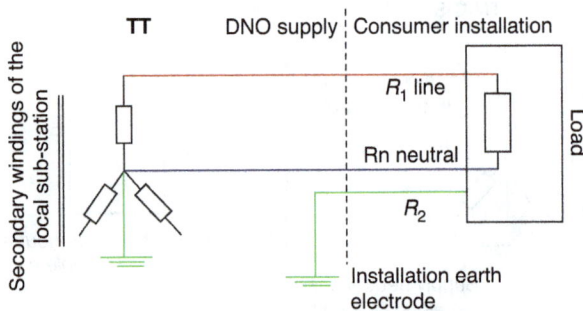

Figure 13.14   Terra-terra (TT) earthing system

13.10   The Earth Fault Loop path

In Section 13.7, we saw that the combination of line and earth conductors feeding an item of electrical equipment takes the form of a loop. This is called the Earth Fault Loop. The impedance of this complete loop is of vital importance because the lower the impedance, the higher the fault current, and the quicker the protective device will operate in the event of a fault.

Note that impedance is to be ascertained, rather than resistance. This is due to the presence of highly inductive windings in the supply transformer. They will possess reactance that has to be taken into account (see Chapter 12).

An earth fault loop impedance test must be carried out for every circuit. The result is then compared with the maximum values allowed for that particular type of circuit. These are to be found in Appendix B of the IET On-Site Guide, as well as Part 4 of BS 7671.

The impedance of the full fault loop path is the sum of two values.

- $R_1 + R_2$ – the total combined resistance of the line and cpc of the circuit.
- Z_e – the impedance of the external part of the loop, the part that belongs to the DNO. The Z_e can be obtained by:
  - Measurement – a test carried out as close to the supply point as possible
  - Calculation – $Z_e = Z_s - (R_1 + R_2)$.
  - Inquiry – the DNO should be able to inform you of the value.

It follows that Z_s can be calculated from the formula:

$$Z_s = Z_e + (R_1 + R_2)$$

where Z_s is the full earth loop impedance; Z_e is the impedance of external part of the earth fault loop; R_1 is the resistance of circuit line conductor; and R_2 is the resistance of cpc.

It is recommended that the Z_s is calculated because the test needed for its measurement is carried out on live circuits. This can be awkward and hazardous. The Z_e test is also a live test but can be carried out with far less risk. $R_1 + R_2$ measurement is carried out on a de-energised "dead" circuit.

Maximum $R_1 + R_2/m$ values for various conductor sizes and line/cpc combinations are given in Appendix I of the IET On-Site Guide.

Example 13.2
What is the Z_s of a lighting circuit if the line has a total resistance of 12.8 mΩ, the cpc has a total resistance of 7.5 mΩ and the Z_e has a value of 0.2 Ω?

$$Z_s = Z_e + (R_1 + R_2) \quad Z_s = 0.2 + (0.0128 + 0.0075) = 0.22\Omega$$

Example 13.3

What is the Z_e of a circuit if the Z_s if 0.77 Ω, the total resistance of the line is 22.5 mΩ, of the cpc, also 22.5 mΩ.

$$Z_s = Z_e + (R_1 + R_2); \quad \text{therefore,} \quad Z_e = Z_s - (R_1 + R_2)$$

$$Z_e = 0.77 - (0.0225 + 0.0225) = 0.73 \ \Omega$$

If the Z_s value exceeds the maximum value given in BS7671 or the IET On-Site Guide, insufficient current will flow to ensure that the protective device operates within its maximum disconnection time. This delay means that the voltage on the metalwork caused by the fault, may exceed 50 V, resulting in electric shock.

The figure of 50 V is derived from the following relationships, found in BS7671.

For TN-S and TN-C-S systems,

$$Z_s \times I_a \leq U_o \times C_{min}$$

where Z_s is the full earth loop impedance; I_a is the current at which a protective device will operate; U_o is the nominal voltage to earth; C_{min} is a factor that takes into account voltage variations caused by external factors. This is given as 0.95 by the Electricity Supply Quality and Continuity Regulations.

For TT systems,

$$R_A \times I\Delta n \leq 50 \text{ V}$$

where R_A is the sum of the resistances of the earth electrode and all earth conductors connecting exposed conductive metalwork in the installation and $I\Delta n$ is the operating current of main fault protection RCD protecting the installation.

13.11   Equipotential bonding

Not only does exposed conductive metalwork have to be earthed, other metalwork at risk of becoming live also needs to be brought to the same potential as the installation earth. This metal is called extraneous metalwork and the earth connections are called equipotential bonding.

Main equipotential bonding

If the main gas and water pipes entering a premises are metal, they should be bonded directly to earth. It is standard practice to include a main earthing terminal (MET) with the mains intake equipment. This is a simple connector block which, unlike the other supply intake equipment, is available to the installer for connection of various earth conductors. The main bonding conductors for the water and gas supply pipes will be connected to the MET. The distance from the pipe to MET should be as short as possible.

Each pipe must have a separate bonding conductor. It is not acceptable to link a single conductor from one to the other. The connection on each pipe must be fitted with a metal label that bears the following legend;

"SAFETY ELECTRICAL CONNECTION. DO NOT REMOVE"

The sizes for main bonding conductors can be found in the IET On-Site Guide where there are tables stating minimum cross-sectional areas for various earthing conductors.

Supplementary equipotential bonding

As well as the main incoming water and gas pipes, other metalwork may be at risk of becoming live in the event of an earth fault. This extraneous metalwork includes:

- Pipes
- Central heating components
- Structural metalwork
- Ductwork

Extraneous metalwork can be connected to earth via:

- Separate conductors to the consumer unit
- Connection to exposed conductive metalwork
- Other extraneous metalwork already bonded to earth

It is preferred to run separate conductors to items of extraneous metalwork; however, if it is more convenient to use a single conductor to bond a number of items such as pipes, the conductor must not be cut. The insulation should be stripped away but the conductor itself should remain intact.

13.12   Summary of formulae used in Chapter 13

Three-Phase

Star

$$U_L = U_p\sqrt{3} \quad U_p = \frac{U_L}{\sqrt{3}}$$

$$I_L = I_p$$

Delta

$$U_L = U_p$$

$$I_L = I_p\sqrt{3} \quad I_p = \frac{I_L}{\sqrt{3}}$$

Earth fault loop impedance

$$Z_s = Z_e + (R_1 + R_2); \quad \text{therefore,} \quad Z_e = Z_s - (R_1 + R_2)$$

where Z_s is the full earth loop impedance; Z_e is the impedance of external part of the earth fault loop; R_1 is the resistance of circuit line conductor; and R_2 is the resistance of cpc.

Requirement for maximum voltage of 50 V on metalwork during an earth fault
For TN-S and TN-C-S systems

$$Z_s \times I_a \leq U_o \times C_{min}$$

where Z_s is the full earth loop impedance; I_a is the current at which a protective device will operate; Uo is the nominal voltage to earth; and C_{min} is a factor that takes into account voltage variations caused by external factors. This is given as 0.95 by the Electricity Supply Quality and Continuity Regulations.

For TT systems

$$R_A \times I\Delta n \leq 50 \text{ V}$$

where RA is the sum of the resistances of the earth electrode and all earth conductors connecting exposed conductive metalwork in the installation and $I\Delta n$ is the operating current of main fault protection RCD protecting the installation.

13.13   Exercises

1. At what voltage is the UK supply generated?
2. What are the two voltages found on the UK supergrid?
3. What is biomass generation?
4. What are the input and output voltages in a local substation?
5. Calculate the line voltage for a star-wound three-phase system if the phase voltage is 11 kV.
6. Calculate the phase voltage for a star-wound system if $U_L = 500$ V.
7. Calculate the phase current and voltage for a delta-wound system if the line current is 83 A and the line voltage is 33 kV.
8. Calculate the line current for a delta wound supply with a 40 A phase current.
9. What should be done if an installation contains both current and pre-2004 wiring colours?
10. What is the maximum current allowed for single-phase supply?
11. Where is the supply earth connected at the local substation?
12. What is an MET?
13. What does the acronym PME stand for and what is it?
14. How is the installation earth connected to the substation earth for each of these earthing systems?
    (a) TT
    (b) TN-S
    (c) TN-C-S
15. Give three items that can be used as an earth electrode.

16. What is the advantage of connecting the installation earth to the neutral at the installation intake?
17. Which conductors are:
    (a) R_1
    (b) R_2
    (c) R_n
18. What is R_1+R_2?
19. Calculate the Z_s of a 15-m long circuit if the combined resistance of R_1 and $R_2 = 0.0027$ Ω/m and $Z_e = 0.28$ Ω.
20. Calculate the length of a circuit if $R_1 + R_2$ is 10.49 mΩ/m, $Z_e = 0.3$ Ω and $Z_s = 0.51$ Ω

13.14 Multiple-choice questions

13M1 Which of these is not an advantage of three-phase?
   (a) Load is shared over the three phase which means smaller conductor sizes
   (b) Three-phase motors have many advantages over single-phase motors, including smaller size, steady torque output, and the ability to self-start.
   (c) It is less reactive and reduces power factor
   (d) Provides both three and single-phase supplies

13M2 What are the insulation colours for three-phase conductors?
   (a) Red, yellow and blue
   (b) Red, white and blue
   (c) Brown, white and black
   (d) Brown black and grey

13M3 What is the first part of a supply that belongs to the consumer?
   (a) The meter
   (b) The main isolator
   (c) The tails
   (d) The main fuse

13M4 Which of these conductor combinations are considered to be live by BS7671?
   (a) Line and neutral
   (b) Line only
   (c) Neutral and earth
   (d) Line and earth

13M5 Which of the following is a high current protective switch which adjustable current settings, used as the main protective device for commercial and industrial installations?
   (a) MCB
   (b) MCCB
   (c) RCB
   (d) RCBO

13M6 What are the two purposes of earthing?
(a) To improve current flow and shield the mains from interference
(b) To provide a negative connection for a supply and for shock protection
(c) To prevent third harmonics and from excessive inductance
(d) To prevent excessive capacitance and reduce thermal constraint in conductors

13M7 What is basic protection?
(a) Protection from shock in the event of earth faults
(b) Protection from shocks in situations where there is an excess of moisture
(c) Protection from shock under no fault conditions
(d) Protection from shock due to exposed conductive parts

13M8 Which of these provides basic protection?
(a) Obstacles and barriers
(b) CPC
(c) Main bonding conductor
(d) Protective devices

13M9 Which of these is an example of exposed conductive metalwork?
(a) Metal trunking
(b) Metal air duct
(c) Pipework connecting a central heating pump
(d) Incoming gas and water pipes

13M10 The main bonding conductor connects
(a) Incoming metallic water and gas pipes to the MET
(b) Incoming water pipe and supply cable sheath to MET
(c) Incoming metallic water and gas pipes to each other
(d) All pipe work to the MET

13M11 What is the name of the conductor that connects non-electrical metalwork to the installation earth?
(a) Main equipotential bonding conductor
(b) Earthing conductor
(c) Supplementary equipotential bonding conductor
(d) Circuit protective conductor

13M12 What is the name of the conductor that connects exposed conductive metalwork to the installation earth?
(a) Main equipotential bonding conductor
(b) Earthing conductor
(c) Supplementary equipotential bonding conductor
(d) Circuit protective conductor

13M13 What is the name of the conductor that connects the installation earth to the MET?
    (a) Main equipotential bonding conductor
    (b) Earthing conductor
    (c) Supplementary equipotential bonding conductor
    (d) Circuit protective conductor

13M14 The impedance of the complete fault loop for a circuit is:
    (a) Z_e
    (b) Z_s
    (c) Z_m
    (d) Z_o

13M15 The maximum voltage that should appear on exposed conductive metal-work in the event of an earth fault is:
    (a) 24V
    (b) 100V
    (c) 50V
    (d) 230V

Chapter 14

Protective devices

If a fault occurs in a circuit or installation, there is usually no time to switch off the supply manually in time to limit any damage it might cause. In the case of an earth fault, there is the added risk of shock.

Automatic switching devices are therefore included in every circuit. Circuit breakers and residual current devices (RCDs) are examples of these protective devices. However, to ensure that they work properly and offer complete protection against fire and shock, the appropriate devices must be installed for a particular circuit.

14.1 Circuit breakers

Circuit breakers provide both overcurrent and shock protection (Figures 14.1 and 14.2).

Overcurrent is current that flows during an electrical fault and is above the rated current of the circuit. There are two types of fault that cause overcurrent:

- Short circuit – an unintended direct or low-resistance connection between
  o Line and neutral
  o The phases of a three-phase circuit

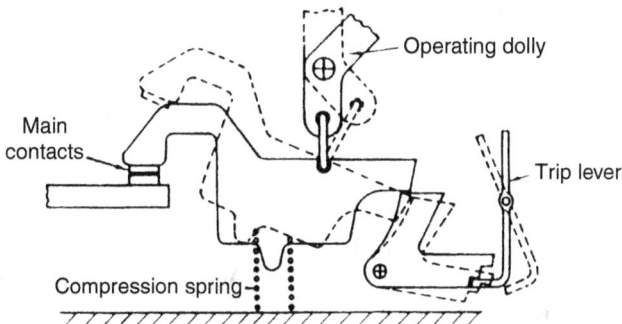

Figure 14.1 Trip mechanism of miniature circuit breaker

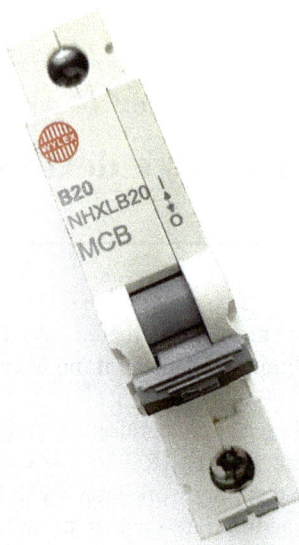

Figure 14.2   Single-phase circuit breaker (by kind permission of Oaklands College, St Albans)

- Overload – high current in an otherwise healthy circuit that flows when the load is too high for the cable that feeds it.

Earth faults are technically short circuits between a live conductor and earth. However, these carry the extra risk of shock, so protection in these cases comes under the heading of shock protection.

The circuit breaker is an automatic switch, which opens when a current flows that is in excess of its rating. Once the cause of the overcurrent has been diagnosed and repaired, the circuit breaker can be switched on again. The contacts of a circuit breaker are closed against spring pressure and held in the closed position by a latch arrangement. A small movement of the latch will release the contacts, which will open quickly under spring pressure to break the circuit.

Circuit breakers have many advantages over fuses, but are more expensive. They are made in a wide range of sizes, from the 6 A to 125 A miniature type for domestic use, up to industrial types capable of switching thousands of amperes.

Thermal tripping – overload

The load current is passed through a small heater, the temperature of which depends on the current it carries. This heater is arranged to warm a bimetal strip, which is made of two different metals securely riveted or welded together along their length. The rate of expansion of the metals is different, so after the strip is warmed it will bend and trip the latch.

The bimetal strip and heater are so arranged that normal currents will not heat the strip to tripping point. If the current increases beyond the rated value, extra power is dissipated by the heater ($P = I^2R$). The bimetal strip temperature is raised to a value that will cause it to bend and trip the latch.

There is always some time delay in the operation of a thermal trip, since the heat produced by the load current must be transferred to the bimetal strip. Thermal tripping is thus best suited to overloads of comparatively long duration.

Magnetic tripping – short circuits and earth faults

The principle used here is the force of attraction that can be set up by the magnetic field of a solenoid coil carrying the load current. At normal currents, the magnetic field is not strong enough to attract the latch, but overload currents operate the latch and trip the main contacts (Figure 14.3).

Magnetic trips are fast acting for heavy overloads, but uncertain in operation for light overloads. The two methods are, therefore, combined to take advantage of the best characteristics of each: Figure 14.3 shows a miniature circuit breaker having combined thermal and magnetic tripping.

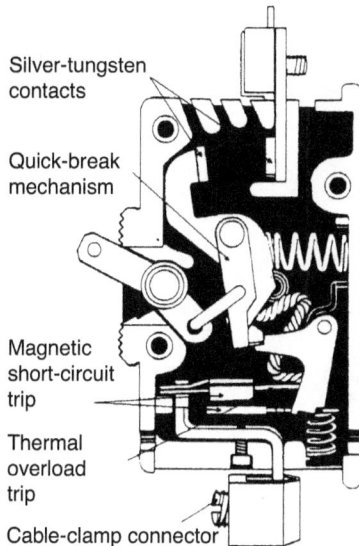

Figure 14.3 General view of miniature circuit breaker

Circuit breaker types and ratings

Circuit breaker types

Circuit breakers are divided into three types:

- Type B – non-inductive loads such as heating appliances. The Type B does not allow a high start-up current to flow without operating.
- Type C – an inductive load current briefly peaks to a value higher than its running current. The Type C will allow a high start-up current to flow for a few seconds before operating.
- Type D – highly inductive loads for which there are large inrush currents at start-up, such as X-ray machines and welding equipment. Type D will allow very high start-up current to flow without operating.

Rating (I_n)

The rating is the maximum current a circuit breaker will allow to flow in a circuit. This current is denoted as In and should always be higher than the design current (I_b) of the load. The design current is the actual current demanded by the load.

The circuit breaker selected should be rated at the next higher current. For example, a 17.5 A load should be protected by a 20 A circuit breaker.

$$I_n > I_b$$

Operating current (I_a)

A circuit breaker will not operate the moment the current rises above its rated value. The IET On-Site Guide gives the overcurrent values that will cause operation of a device.

These are:

- Type B – 3 to 5 × I_n
- Type C – 5 to 10 × I_n
- Type D – 10 to 20 × I_n

Example 14.1

What are the minimum and maximum operating currents of the following circuit breakers?

(a) 6 A Type C
(b) 32 A Type B
(c) 40 A Type D

(a) *Type C*
    Minimum 5 × I_n 5 × 5 = 25 A Maximum 10 × I_n 10 × 5 = 50 A
(b) *Type B*
    Minimum 3 × I_n 32 × 3 = 96 A Maximum 5 × I_n 32 × 5 = 160 A

(c) *Type D*
   Minimum $3 \times I_n$ 40 $\times$ 10 = 400 A Maximum 20 $\times I_n$ 40 $\times$ 20 = 800 A

These operating currents may seem high, but a large current will flow when a short circuit or earth fault occurs. The key is to enable the fault current to rise to the required level as quickly as possible so that the circuit breaker operates within the disconnection times laid down in BS7671. These are:

400 V/230 V AC TN-S and TN-C-S systems
(see Chapter 13 for more details on these systems)

- 0.4 s
  - o Circuits feeding loads ≤32 A
  - o All circuits feeding socket outlets rated at ≤63 A

- 5 s
  - o Circuits feeding loads >32 A
  - o Sub-mains

400 V/230 V AC TT systems
(see Chapter 13 for more details on this system)

- 0.1 s
  - o Circuits feeding loads ≤ 32 A
  - o All circuits feeding socket outlets rated at ≤ 63 A

- 2 s
  - o Circuits feeding loads ≤ 32 A
  - o Sub-mains

A particular circuit breaker is chosen to offer the best protection for the circuit concerned. BS 7671 provides graphs of time/current characteristics for all three types of circuit breaker.

Achieving the current required to operate the protective device within its disconnection time depends on a low earth fault loop impedance. This is discussed more fully in Chapter 13, but put simply, it is the impedance of the path travelled by an earth fault current from the fault to the supply transformer and the path travelled by the prospective fault current back to the point of fault.

Breaking capacity

As previously stated, high values of fault current will flow when a short circuit or earth fault occurs. The protective device needs to endure these high currents and still operate effectively. For this reason, circuit breakers are constructed with a range of breaking capacities, e.g. 4 kA, 6 kA and 12 kA. When carrying out test and inspection, the prospective fault current for an installation is ascertained (by measurement, calculation or inquiry to the DNO) to confirm that the protective devices installed are of the correct breaking capacity.

Wireless-operated circuit breakers

A further development in circuit breaker technology is wireless-operated devices. The purpose of the remote control is to switch circuit breakers off as part of the safe isolation process. Among other advantages:

- It makes the identification of the circuit origin physically easier in that the electrician no longer needs to walk to and from the consumer unit/distribution board when using the trial-and-error method of locating the associated protective device.
- It also facilitates emergency isolation in the event of shock or other dangerous occurrences.

14.2   Residual-current device (RCD)

We saw in Chapter 13 that the earth-fault-loop impedance should be low enough to ensure that enough current flows in the event of a fault to operate the circuit breaker or blow the fuse. In some installations, particularly those which require an earth electrode, the measured impedance may be too high to ensure the timely operation of the circuit breaker or fuse and a residual-current device (RCD) must be used (Figures 14.4 and 14.5).

Figure 14.4   Typical RCD arrangement in a consumer unit (by kind permission of Oaklands College, St Albans)

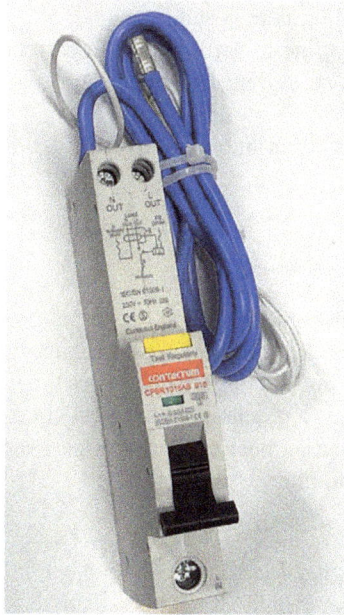

Figure 14.5 Typical RCBO (by kind permission of Oaklands College, St Albans)

A simplified arrangement is shown in Figure 14.6. The heart of the device is a magnetic core, which is wound with two main current-carrying coils, each with the same number of turns. These are connected so that the phase current in one coil provides equal and opposite ampere-turns (see Section 6.4) to the neutral current in the other coil, as long as these two currents are equal the device will not operate. A test button is provided, which simulates a fault and is used to confirm that the device is functioning correctly. The symbol for RCD rating is *I∆n*. Operation of the RCD is as follows:

1. Normal conditions – in compliance with Kirrchoff's Law, the outgoing current through the line conductor and incoming current through the neutral are equal.

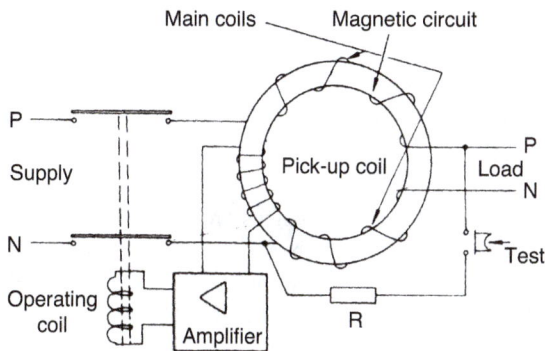

Figure 14.6 Simplified arrangement for residual-current device

2. Magnetic flux in the iron core is stable.
3. Earth fault occurs, e.g. line-to-earth fault – all current directly to earth through line winding. No return current through neutral winding – flux in iron core changes.
4. Flux change induces current into search winding, which operates the solenoid and opens switch.

RCD functions

Additional Protection

RCDs of 30 mA can be utilised to provide additional protection. This is, in effect, an extra, close layer of shock protection. The disconnection time for an additional protection RCD is ≤ 300 ms, although much faster times are expected.

Additional protection is a requirement for most circuits and consumer units are designed to accommodate a number of these devices, each one of which will protect a single circuit or a number of circuits.

Fault protection

These generally have a higher $I\Delta n$ and provide protection from shock to a complete installation, or part of an installation, rather than to an individual circuit. For example, if an installation is protected by a TT earth system, which uses an earth electrode, the whole installation must be protected by a fault protection RCD. Fault protection can be rated from 30 mA to 300 mA.

Fire protection

About 300 mA RCDs can be used for fire protection. This may seem an odd use for a device primarily designed to provide protection from electric shock, but its rapid action makes it an ideal fire protection device. When a cable burns and melts, it will cause a short circuit between the live conductors and the circuit protective conductor (the circuit earth), which will result in RCD operation.

S-Type

If the installation, or part of the installation, is protected by a fault protection RCD and further down the line there are lower rated RCDs protecting individual circuits, there is a chance that the main RCD, rather than the additional protection device, will operate in the event of a fault. This can be inconvenient because the whole installation will be switched off. The S-type RCD has a time delay, which gives the additional protection device time to operate and isolate only the faulty circuit.

RCD types

Modern electrical equipment and appliances are increasingly complex in terms of operation and in their reliance on both AC and DC supplies. Electronic components, including LED screens, are a common part of many household appliances. These mixed voltage types can cause problems with RCD operation. Specialised

RCDs have been developed to counter these issues and continue to provide close protection for complex circuits and loads.

Type A

Intended for both AC residual current and for residual pulsating DC up to 6 mA. Type A RCDs protect circuits that include equipment fitted with electronic components. For example:

• Inverters
• LED drivers
• Lighting dimmers
• Induction hobs
• Electric vehicle charging equipment with smooth residual DC current less than 6 mA

Type AC

Designed for AC residual current in circuits without any electronic components, the Type AC RCD, or General Type, is the most common type installed in domestic dwellings. They protect equipment that is:

• Resistive
• Capacitive
• Inductive

Examples of suitable circuits:

• Electric showers
• Electric heaters
• Oven
• Tungsten lighting

Type F

Type F RCDs are used for appliances and equipment that incorporate frequency speed control (see Book 2 for more detailed description of frequency control).

Examples of equipment include:

• Some Class I power tools
• Air conditioning controllers with variable speed drives
• Washing machines
• Dishwashers*
• Tumble driers*
  (*Driven by synchronous motors)

  Type F devices are also suitable for Type AC and Type A applications.

Type B

Type B RCDs are used for single- and three-phase equipment. They are intended for DC circuits with a leakage current in excess of 6 mA. Examples of equipment include:

- Inverters
- Uninterruptible power supplies (UPS)
- Photovoltaic systems
- Lifts
- Escalators
- Welding equipment
- Electric vehicle charging equipment with smooth residual DC that is greater than 6 mA

Type B devices are also suitable for Type AC, Type A and Type F applications.

To prevent the possibility of the potential difference between conductive parts and earth becoming too high for safety, this type of protection can only be used when the operating current in amperes multiplied by the earth-loop impedance in ohms does not exceed 50 V.

$$U = IZ \leq 50 \text{ V}$$

14.3   Residual circuit – break on overcurrent devices (RCBO)

Because of the increased use of RCDs a combined protective device that incorporates both functions is now in wide use. This is called the Residual Circuit – Break on Overcurrent device, or RCBO. There are many advantages of this combined device.

RCBOs protect individual circuits whereas an RCD often protects a group of circuits. This means that in the event of an earth fault with its associated risk of electric shock, only the faulty circuit will be isolated.

Because they include the RCD function, RCBOs should be subject to the RCD tests as laid out in IET Guidance Note 3 (see Chapter 14 in Volume 2). They also feature a test button that should be operated six-monthly to check that the device is fault-free.

The overcurrent protection function of an RCBO is, essentially, a circuit breaker. As with individual circuit breakers, they are manufactured as Types B, C and D. They must also operate within the disconnection times stated in BS7671 and circuits protected by these devices must conform to the maximum Z_s values also stated in BS7671.

14.4   Fuses

A fuse is the simplest overcurrent protective device. It consists of a thin wire or metal tape placed in series with the circuit to be protected so that it carries the circuit current. The wire is thick enough to carry normal load current without overheating, but if the current exceeds the normal value, the fuse wire will melt, breaking the circuit (Figure 14.7).

The thickness of the wire, or fuse element, is of obvious importance. If it is too thin, its resistance will be high, and the power dissipated in it by the rated circuit current will raise its temperature to melting point. If it is too thick, its low resistance will dissipate little power, so that it may not melt even if the current becomes large enough to damage the circuit conductors or apparatus.

Similarly, the fuse enclosure is important. If the fuse element is open and unshrouded, it will quickly dissipate heat and will require a large current to melt it. If enclosed so that the heat cannot escape easily, it will melt at a lower current.

Fuses are no longer in general use as circuit protection devices and have been widely replaced by circuit breakers and RCBOs. However, the main protective device for most domestic installations will still be a fuse, usually 100 A. The main uses for fuses nowadays are:

- Protection of flexible cable that connects electrical equipment to the supply. A plug top should be equipped with a BS 1362 fuse rated at either 3 A or 13 A.
- Protection for electrical equipment fed off a higher rated circuit such as a ring final circuit. For example, a spur from a ring final circuit intended to feed a small water heater. A fused connection unit equipped with a BS 1362 fuse should be installed as the isolator and connection point for the equipment.
- Internal fuses for electrical and electronic equipment.

Figure 14.7   Typical fuses (by kind permission of Oaklands College, St Albans)

Fuses are designed to operate at different speeds, according to their intended use. These are:

- Ultra-rapid fuses – react almost instantaneously and protect against short circuits in electronic circuits. They are also known as
  o Very Fast Acting
  o Super Rapid
  o High-Speed fuses

- Fast-Acting Fuses – a general-purpose fuse used to protect cables and larger components from overcurrent faults. They are also known as:
  o Fast Blow
  o F-Type fuses

- Slow-Acting fuses - include a delay mechanism that will briefly allow electrical surges to pass through the circuit without blowing. Dual-Element versions provide a higher performance delay. They are also known as
  o Slow Blow
  o Time Delay fuses

High rupture capacity (HRC) fuses

An HRC, or BS 88, fuse has its element enclosed in a cartridge of heatproof material, the cartridge being packed with chemically graded quartz to prevent the formation of an arc, so there is no fire risk, owing to the enclosure of the element. This type of fuse is:

- Non-ageing
- Fast operating
- Accurate in that it operates at a definite current for a given rating
- Capable of breaking very heavy current.

Different current ratings are usually made in different sizes, which makes it impossible to fit a fuse of the incorrect rating. The construction of a typical HRC fuse is shown in Figure 14.8.

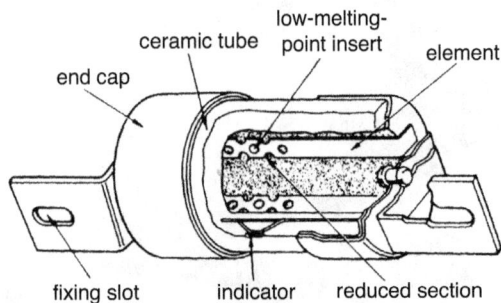

Figure 14.8   High-breaking-capacity fuse

HRC fuses are in wide use, particularly as the main protective device for domestic premises.

General enclosed fuse type designs

- BS 88 designs – enclosed fuses such as BS 88 types are manufactured in a variety of designs. Many are fitted with blades or brackets with which they can be clipped or bolted into place.
- Bottle or diazed fuse – shaped like a miniature bottle, features an indicator which signals that the fuse has blown. Each rating is of a different physical size, making it impossible to install the incorrect fuse for a particular circuit or load.
- Hole-mounted tag fuses – used for semiconductor protection, can be screwed into place.
- Thermal Fuses – used in tumble dryers, coffee makers and other heat-producing accessories, a thermal fuse will blow when the temperature exceeds its rated holding value.
- Auto fuses – blade-type fuses that protect the electrical and electronic circuits in cars.

Semi-enclosed fuses

No longer installed, semi-enclosed fuses are still in place in many existing installations and are worth examining. The fuse element consists of a wire that is connected to two screw terminals on the fuse carrier. The wire is usually made of an alloy containing 63% tin and 37% lead. Semi-enclosed fuses have the advantages of low cost and simplicity, but have many disadvantages, including

- Ageing: oxidation of the fuse elements often leads to failure of the fuse at rated current.
- Time delay in operation: this is significant when compared with high-breaking capacity (HRC) fuses.
- Variations in fusing current owing to the use of elements of different composition and differing enclosures.
- Low rupturing capacity: in the event of a severe fault, the current may vaporise the element and continue to flow in the form of an arc across the fuse terminals.
- Fire risk: the element becomes white hot as it operates.
- A incorrect fuse rating can be fitted, deliberately or by accident, by connecting a fuse wire, or element, of the wrong size. A 5 A circuit protected by 60 A fuse wire in a 5A carrier results in an obvious hazard.

BS 7671 requires that the rating of a semi-enclosed fuse shall not exceed 0.725 times the current-carrying capacity of the lowest-rated conductor protected. This means that larger cables may be necessary where semi-enclosed fuses are used to provide overcurrent protection.

DC fuses

Different fuses must be used for AC and DC protection. There are a number of reasons why a dedicated DC fuse should be used for DC circuits and loads.

- Arcing – because of the steady current flow in a DC system, arcing takes place when a circuit is broken either by a switch or by a fuse. A DC fuse is designed to handle this problem.
- Voltages – AC fuses can be used for much higher voltages than DC fuses.

Fusing factor

The current rating of a fuse is the current it will carry continuously without deteriorating. The minimum fusing current is the current that will cause the fuse to operate under given conditions in a given time.

$$\text{Fusing factor} = \frac{\text{Rated minimum fusing current}}{\text{Current rating}}$$

It follows that the fusing factor must exceed one. The closer it is to one, however, the less likely it is that a fault current will fail to operate the fuse. Put simply, the lower the fusing factor, the better the fuse.

The time taken for a fuse to operate is a function of the current carried. BS 7671 provides graphs of time/current characteristics for all types of fuse.

14.5   Discrimination

A normal installation has a number of protective devices connected in series. For example, the main MCCB for a factory may be rated at 200 A, with a 60 A circuit breaker feeding a distribution board that, in turn, houses 16 A circuit breakers or RCBOs protecting individual subcircuits (Figure 14.9).

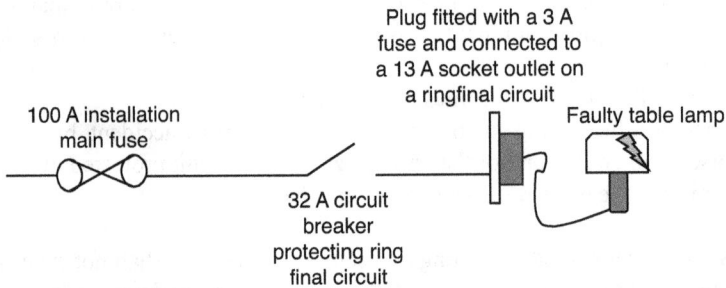

Figure 14.9   Good discrimination

Obviously, a fault on the final subcircuit should cause the 16 A device to blow. If either the 60 A or the 200 A devices blow, this will show a lack of discrimination on the part of the fuses.

Some circuit breakers and fuses are particularly fast in operation. Care must be taken to ensure that:

- That such devices are not used to protect circuits in which much slower-operating but lower-rated fuses are used.
- The main fuse or circuit breaker doesn't operate before the device in the sub-circuit, putting a much wider area than necessary out of action.

Either of these eventualities would show bad discrimination. BS 7671 states that there must be an overcurrent protective device at every point at which con-ductor size is reduced.

14.6   Surge protection devices (SPD)

Whilst transient overvoltage events have always been a problem and damaging to electrical equipment, the wide use of solid-state, electronic circuitry has heightened the threat posed by these supply disturbances. High voltages can cause severe damage to semi-conductor components (Figure 14.10).

Voltage surges can be caused by:

- Short circuits and earth faults on the HV side of a supply transformer.
- Lightning strikes on supply equipment.

Figure 14.10   Typical surge protection device (by kind permission of Oaklands College, St Albans)

The surge protection device is now in wide use. Designed to fit in the ways of a consumer unit or distribution board in the same manner as a circuit breaker, SPDs can be installed both at the source of the circuit and also adjacent to the vulnerable equipment.

- Type 1 SPD installed at the origin, e.g. main distribution board, can handle very high voltage surges caused by lightning strikes. Spark gap technology is employed to short high currents to earth.
- Type 2 SPD installed at sub-distribution boards. They protect from both lightning and HV switching faults. They employ metal oxide or MOV to divert the current through the device in the form of a short circuit. Combined Type 1 & 2 SPDs are available and are usually installed in consumer units.
- Type 3 SPD installed close to the protected load and is designed to safeguard sensitive electronic equipment. They must only be installed as a supplement to a Type 2 SPD.

The basic principle of SPDs is based on the relationship between voltage, resistance and current (Figure 14.11). By creating a low resistance or impedance path through itself, an SPD will cause high current and, therefore, low voltage to be present. This is achieved using:

- **Sintered** zinc oxide – the multiple joins between the particles form a network of series and parallel paths through which the excess current it dispersed. These connections have high resistance at normal low voltage, low resistance at high voltage.
- A series of reverse bias diodes – avalanche breakdown.
- Gas discharge tubes – spark gap types. Two electrodes, separated by a gap, are positioned in the tube. Under normal conditions the current cannot flow between them, but with increased voltage the gas impedance is lowered enough for the current to arc across.

KEY TERM: Sintering – A process whereby particles of a material such as zinc are heated to a temperature high enough to bind them together into a solid, but not high enough to melt and weld them together

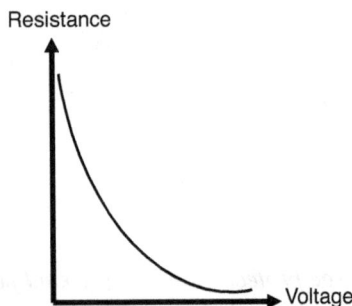

Figure 14.11   Relationship between voltage and resistance

Under normal conditions, the high resistance within the device prevents current flowing across its terminals.

Under voltage surge conditions, the resistance/impedance effectively forms a short circuit through the device that allows excess current to flow through it. Because voltage and current are inversely proportional, the voltage value is reduced by the high current flow.

14.7   Arc fault detection devices (AFDD)

Loose connections and damage to cables and equipment can result in electrical arcing. While small, an arc can reach high temperatures and cause overheating and fire. It can also lay down a carbon path across insulated parts that could result in a short circuit or earth fault. The arc fault detection device is designed to detect the presence of an arc in a circuit and operate to isolate the circuit before excessive damage or a fire occurs (Figure 14.12).

Common causes are:

- Kinks in a cable
- Breaks or partial breaks in a cable
- Incorrectly stripped cables
- Incorrect bonding
- Lose connections
- Defective plugs and worn insets in socket outlets that cause the plug-and-socket connection to be loose
- Rodent damage

Figure 14.12   Typical AFDD (by kind permission of Oaklands College, St Albans)

Arc faults are either:

- Series – broken or loose connections, damaged cables
- Parallel – damage that causes arcing between cable cores, i.e. line-to-neutral or live-to-earth

AFDDs are mandatory for circuits containing socket outlets rated at ≤ 32 A in the following premises:

- Higher Risk Residential Buildings (HRRB) – buildings over 18 m high or over six floors
- Houses in Multiple Occupation (HMO)
- Care Homes
- Purpose-Built Student Accommodation

Like SPDs, AFDDs are manufactured so as to fit into the ways in a consumer unit or distribution board.

Using a microprocessor, the device monitors the AC waveform, looking for any irregularities that could be a sign of arcing. One of the key components used in some AFDDs is the Rogowski coil. This is a toroidal, conductive coil placed around a conductor. When current flows through the conductor, its resulting magnetic field induces current into the coil. Rogowski coils are able to detect changes in the current rapidly, including the current changes caused by an arc fault. The signal is interpreted through the device's microprocessor that is then connected to a low-current solenoid that operates the device switch.

SPDs and AFDDs have increased the number of protective devices required for an installation, and this means that consumer units and distribution boards need to be designed with more capacity to host these devices. However, just as RCDs and circuit breakers have been combined into single devices called RCBOs, AFDDs and SPDs can be manufactured so that they are also integrated into a single device.

14.8   Multiple-choice questions

14M1 What is an overcurrent in an otherwise healthy circuit is called?
    (a) Short circuit
    (b) Voltage surge
    (c) Overload
    (d) Earth fault

14M2 What is the additional hazard arising from an earth fault?
    (a) Shock
    (b) Fire
    (c) Overheating
    (d) Arc faults

14M3 What is the mechanism inside a circuit breaker that causes its operation under short circuit conditions?
(a) A solenoid
(b) A bimetal strip
(c) A sintered layer
(d) A multi-diode current cascade

14M4 What is the maximum fault current allowed for the operation of a 45 A Type C circuit breaker to operate under fault conditions?
(a) A
(b) B
(c) C
(d) D

14M5 Which circuit breaker Type is suitable for inductive circuits?
(a) A
(b) B
(c) C
(d) D

14M6 What is I_n?
(a) The actual current used by the load
(b) The nominal current flowing in a circuit
(c) The current that will cause effective operation of the device
(d) The rating of a protective device

14M7 What is I_b?
(a) The actual current used by the load
(b) The nominal current flowing in a circuit
(c) The current that will cause effective operation of the device
(d) The rating of a protective device

14M8 What is I_a?
(a) The actual current used by the load
(b) The nominal current flowing in a circuit
(c) The current that will cause effective operation of the device
(d) The rating of a protective device

14M9 What is breaking capacity when used as a circuit breaker rating?
(a) The amount of fault current that enables it to meet its disconnection time
(b) The amount of fault current it can withstand and still work effectively
(c) The amount of fault current that will cause its operation
(d) The amount of fault current that will cause it to malfunction

14M10 What is the maximum disconnection time for a circuit feeding a 32 A socket outlet if it is protected by a TN-S earthing system?
(a) 0.5 s
(b) 2 s
(c) 0.4 s
(d) 5 s

14M11 What is the maximum disconnection time for a circuit feeding a 6 A load if it is protected by a TT earthing system?
(a) 0.2 s
(b) 2 s
(c) 4 s
(d) 0.5 s

14M12 What is the rating that allows the use of an RCD for additional protection?
(a) 300 mA
(b) 600 mA
(c) 16 mA
(d) 30 mA

14M13 What is the maximum rating for a fire protection RCD?
(a) 100 mA
(b) 30 mA
(c) 600 mA
(d) 300 mA

14M14 What does RCBO stand for?
(a) Residual current – bimetal overload
(b) Residual current – break on overcurrent
(c) Reactive current – break on overheat
(d) Reactive current – bimetal overload

14M15 Which type of RCD has a time delay function?
(a) A
(b) S
(c) F
(d) B

14M16 Which type of RCD is suitable for electric vehicle charging points?
(a) A
(b) S
(c) F
(d) B

14M17 What does HRC stand for in relation to fuses?
(a) High Rated Capacity
(b) High Rupture Capacity
(c) High Reactance Capacity
(d) High Regulating Capacity

14M18 Why is it important to protect against transient overvoltage?
    (a) To decrease reactance in circuits
    (b) To reduce wear and tear on mains switch terminals
    (c) To improve Z_s
    (d) To prevent damage to electronic circuitry and equipment

14M19 Which type of Surge Protection Device should be installed adjacent to the equipment it protects?
    (a) 1
    (b) 2
    (c) 3
    (d) 4

14M20 What is an AFDD?
    (a) Arc Fault Detection Device
    (b) Arc Finding Detection Drive
    (c) Arc Fault Direction Divertor
    (d) Arc Fault Detection and Disconnector

Chapter 15

Sub-circuits

15.1   Introduction

Electrical installations can vary widely in scope and complexity, from, for example, a single lighting point and socket outlet in a garden shed to the complete installation of a new hospital or school. However, all installations follow the same basic principles to ensure safety from fire and shock. The installation begins at the meter tails (see Chapter 13) that connect the supply to the distribution or consumer unit. In larger installations the necessary switchgear will be enclosed in a panel.

Lighting and equipment in the installation are fed from the consumer unit/ distribution board. BS 7671 requires the installation to be broken down into separate circuits for the following reasons:

- To prevent danger in the event of a fault. The fault current will be no higher than that required to operate a circuit's protective device
- To enable part of the installation to be isolated for maintenance or fault diagnosis without affecting the whole installation
- To prevent a fault on one circuit from causing the loss of the power for the whole installation

A suitably rated protective device must be installed at the source of each circuit (see Chapter 14).

15.2   Consumer unit/distribution board

There is no clear statement of the difference between a consumer unit and a distribution board. They both serve the same purpose, which is to supply the circuits in the installation. To avoid confusion, the two names will be applied as follows in the remainder of this chapter.

- Consumer unit – a single-phase board used in individual domestic properties
- Distribution board – a single- or three-phase board used in commercial and industrial installations. Not only to feed individual circuits but also sub-mains that supply other distribution boards throughout the premises

All consumer units and distribution boards must be manufactured from non-flammable materials, usually metal. There are many non-metal boards (mostly plastic) still in use, however, but there is no requirement to change these retrospectively.

A typical consumer unit (Figure 15.1) consists of:

- A main **double-pole** switch that can be used to isolate the entire installation
- A busbar to which the protective devices are connected. This busbar will be insulated or concealed to minimise the risk of electric shock
- A neutral bar to which all the circuit neutrals are connected. There may be more than one neutral bar, each dedicated to circuits fed via the RCDs in the board. It is important that the neutral for each circuit is connected to the bar that corresponds with its circuit RCD
- An earth bar into which all the circuit protective conductors are connected.
- Ways for the protective devices, which will include:
  - Circuit breakers
  - Surge Protection Devices (SPDs)
  - ArchFault Detection Devices (AFDDs)
  - Combined devices (e.g. RCBO, RCBO/SPD)

Distribution boards will have the same components. They may also include three bus bars, one for each phase if it is intended for a three-phase and neutral installation. The board will be manufactured to accept both single-phase and three-phase devices. Each way can be utilised by:

- Three single-phase devices
- A single three-phase one

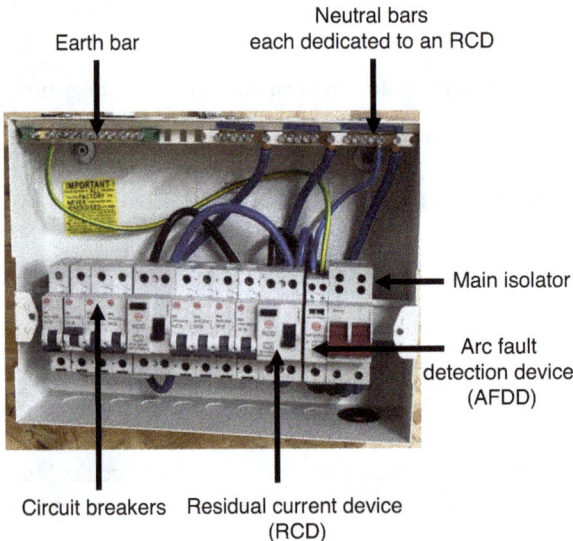

Figure 15.1   Consumer unit (by kind permission of Oaklands College, St Albans)

KEY TERM: Double-pole – A switch that will make and break both line and neutral conductors. A triple-pole device switches all three phases of a three-phase system, a four-pole, all three phases and the neutral.

15.3   The structure of a practical circuit

No matter how complex a circuit is in terms of switching and controls, it will always feature certain components. Let's take a basic circuit, the feed to an electric heater, as an example. This type of circuit, a straightforward load fed by a single line and neutral conductor, and a CPC is called a radial circuit. The three-phase version would, of course, contain three line conductors, a neutral and a CPC.

Source

The circuit begins at the consumer unit: the neutral connected to the appropriate neutral bar, the CPC to the earth bar and the line conductor to the protective device. It is good practice to connect the CPC and neutral to the corresponding terminal in their respective bars. For example, if the circuit breaker is located in the third way counting from the main switch, then the CPC and neutral should be connected to the third terminal in their respective bars.

Cables

Cables provide the connection between the supply and the load. Types of cable and wiring systems are explored in detail in Chapter 16. It is important that the correct cable type and wiring system are used. Cables must be protected from mechanical damage either by a toughened outer sheath or by some form of containment such as conduit and trunking.

Cables must also be sized so that they can carry the required current. The Design Chapter in Volume 2 describes the full process for cable selection. This is, however, an important part of electrical design and safety. An overloaded cable will overheat and is a fire hazard.

Circuit protective conductor

All exposed conductive metalwork should be at the same potential as earth. To achieve this, an earth conductor, known as the circuit protective conductor (CPC) is provided. This can already be included within a multicore cable. In the case of single cables run in conduit or trunking, it will be necessary to add a CPC to the circuit cable loom.

Isolator

The supply to the heater will be connected first to a local isolator (Figure 15.2). This is a double-pole switch designed to completely sever the electrical supply if repair or maintenance work is to be carried out safely on the load. For a three-phase circuit, this would be a three- or four-pole isolator (depending on the presence of a neutral). Note, however, that an isolator should only open and close live conductors, it must NEVER open the CPC.

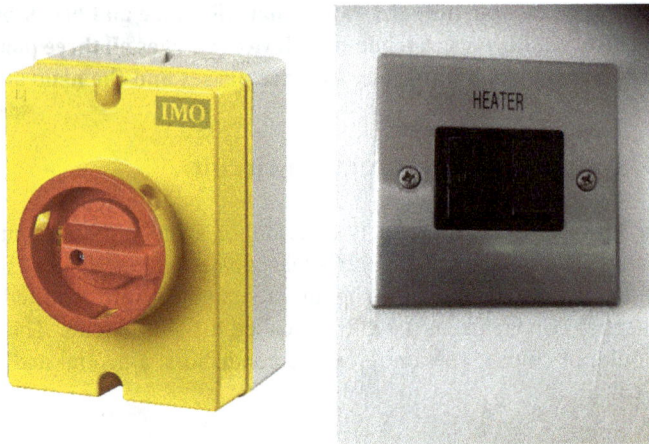

Figure 15.2　Typical isolator used for commercial and industrial installations and domestic heater isolator (by kind permission of IMO Technology)

While an isolator is intended to be operated off-load, that is, when the equipment it protects is not functioning, it must also be capable of operating under full-load conditions without suffering damage.

Some isolators are fitted with combinations of switches, fuse carriers and flex outlets. These types of isolators are often called spurs, or spur outlets. Although the circuit is protected by the device at the consumer unit, a close protection fuse may be required to achieve diversity (see Chapter 14).

Ideally, an isolator should be located adjacent to its associated equipment; however, this is not always practicable, for example, in a bathroom shower or heater. While pull-cord isolators are acceptable for use in a bathroom, a conventional type can only be used if it is installed outside the bathroom itself. In this case the isolator must be labelled.

The perfect isolator is the plug and socket outlet. Once a plug has been removed, there is no possibility of any current getting through to the load.

Isolators are used for two types of maintenance.

Mechanical maintenance

Maintenance and repair are carried out on the equipment itself, such as changing mechanical parts or even replacement of the equipment. The supply requires isolation to prevent accidental start-up, which could result in an accident.

Electrical maintenance

Inspection and testing, fault diagnosis and repair or upgrades of existing accessories and equipment qualify as electrical maintenance. Chapter 2 describes the safe isolation procedure necessary to ensure electrical maintenance can be carried out safely. However, it may not be practicable to isolate an entire circuit, so equipment

Figure 15.3   Examples of functional switches: light switch and portable heater switch

can be made safe using a local isolator. If the whole circuit needs to be de-energised, then the protective device will act as an isolator.

Functional switch

A functional switch is designed to be opened and closed as part of the normal operation of the circuit. It is not intended to make a circuit safe for maintenance or repair work because, although it will stop the circuit's or equipment's function, it will not isolate its supply. Examples of functional switches (Figure 15.3) are:

- Light switch
- Buttons controlling hot plates on a hob
- On-off switch on a television
- Contactor starter for an electric motor
- Thermostat for a heating system

These switches are manufactured for operation under full-load conditions. The current rating of a functional switch is important because of this. For example, the maximum rating for light switches is 10 A. This means that lighting circuits should be designed so as not to exceed this rating, or if unavoidable, high-current lighting circuits should be operated via a contactor, with the light switch switching the coil rather than the load itself.

Discharge lighting (fluorescent lighting, for example) experiences a high current peak at switch-on. Functional switches used for discharge lighting must be capable of withstanding $1.8\times$ the rated current of the circuit.

Emergency switching

While our heater does not require any external emergency switching, there are certain loads that will require a means of rapid isolation for safety. The heater may feature an internal switching device that will operate if it threatens to overheat. Electric motor contactors include a set of overloads, which will operate if the motor experiences a mechanical fault, jams and begins to overheat. The protective device(s) associated with a circuit also act as automatic emergency switches in the event of a fault.

An example of a manually operated emergency switch is a stop button adjacent to a machine, which ensures that it can be stopped quickly in the event of an accident.

Fireman's switch

A fireman's switch is a form of emergency switch. It is located on the outside of a building and intended to allow the fire service to isolate a high-voltage lighting system in the event of a fire. High voltage is defined, in this case, as:

$$\geq 1000 \text{ V AC} \geq 1500 \text{ V DC}$$

A fireman's switch is mounted at a height, which prohibits inadvertent use by the general public. It is operated by a pole carried in fire engines for that purpose.

Note: BS 7671 contains a table of switching devices and indicates whether these can be used for isolation, emergency or functional switching.

Load

The point of the circuit. It can be connected directly into the main installation wiring, for example a luminaire, or it will be connected via a flexible cable or flexible conduit. In most cases, the load is not considered to be part of the fixed wiring of the installation. Lighting is an exception to this rule. The information plate on most electrical equipment will state:

- Voltage
- Frequency
- Power factor (if an inductive device such as an electric motor – see Volume 2)
- Power

The actual current required by the equipment, the design current I_b, can be calculated using the formula:

$$I_b = \frac{P}{V}$$

15.4   Lighting circuits

Extra-low-voltage LED lighting has revolutionised lighting installations but has brought its own challenges. It is rapidly replacing conventional lighting types and lends itself to remote switching and modular installation methods. However, conventional lighting circuits are still very much in evidence, and it is important for the electrician to know and understand how these are wired.

First, we will consider basic lighting circuits. There are generally two ways of wiring these circuits, multi-core cables such as twin-and-earth or singles cables. These necessitate a different approach.

It is good practice to separate lighting into more than one circuit so that a lighting circuit fault does not plunge the entire premises into darkness. Also, the maximum rating for most types of light switch is 10 A.

Basic lighting systems

One-way lighting

This is the simplest lighting control circuit. A single switch controls one or more lights (Figures 15.4 and 15.5).

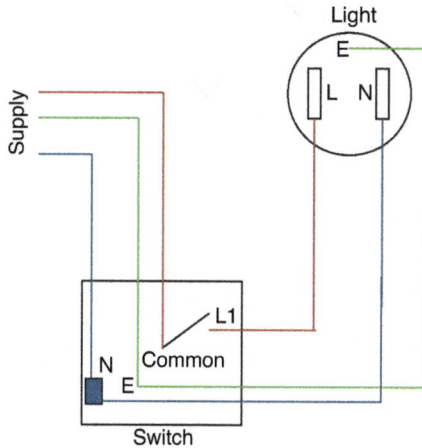

Figure 15.4   One-way light and switch wired in singles. Note: neutral at switch now required – see Digital Lighting.

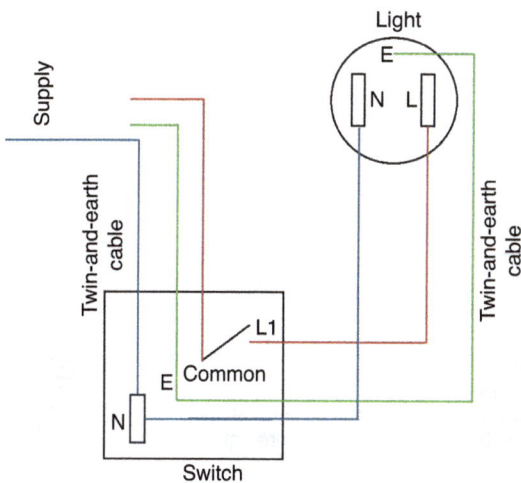

Figure 15.5   One-way light wired in multicore (twin-and-earth) cable

Two-way lighting

A more complex system is one that enables you to switch on the lights with one switch and then switch them off at another, and vice versa. The important principle is that the state of the lights should change with the single operation of either switch. The core of this type of system are the strappers that link the two switches. An example of this type of system would be a room with two entrances/exits or lights on a stairway with a switch at the top and one at the bottom (Figures 15.6 and 15.7).

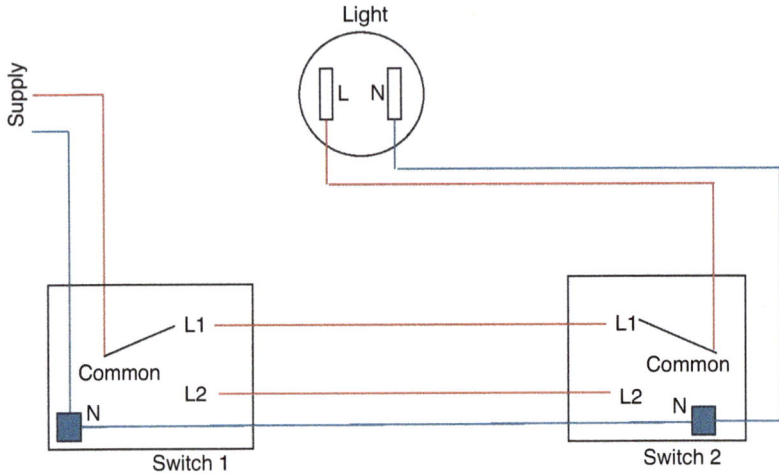

Figure 15.6   Two-way lighting wired in singles (earth omitted for clarity)

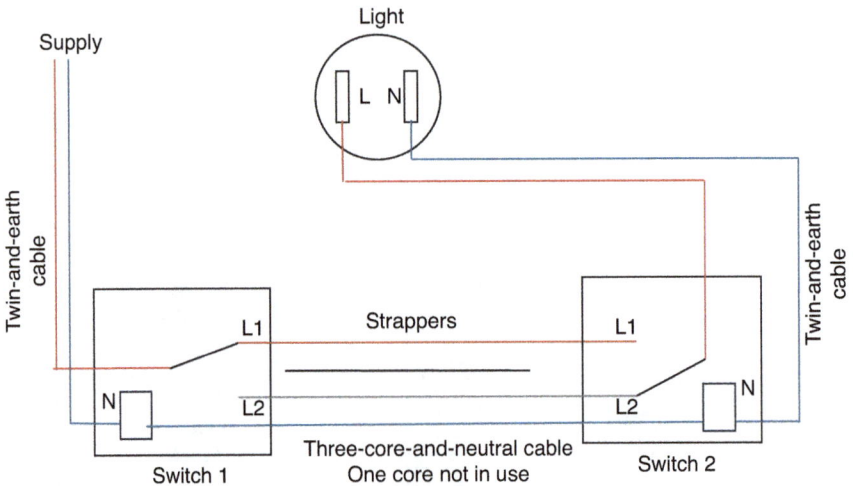

Figure 15.7   Two-way lighting wired in multicore cables (earth omitted for clarity). The spare core of the three-core cable could also be used as the linking neutral between the two switches, provided it is clearly identified as a neutral with blue sleeving.

Two-way and intermediate

This is a more complex version of two-way switching in which more switches are added in between the first and last switch. The number of intermediate switches is unlimited. An example of intermediate switching would be a hotel corridor, e.g. a two-way system to control the lights from either end of the corridor, with inter-mediate switches adjacent to the doors of each room (Figures 15.8 and 15.9).

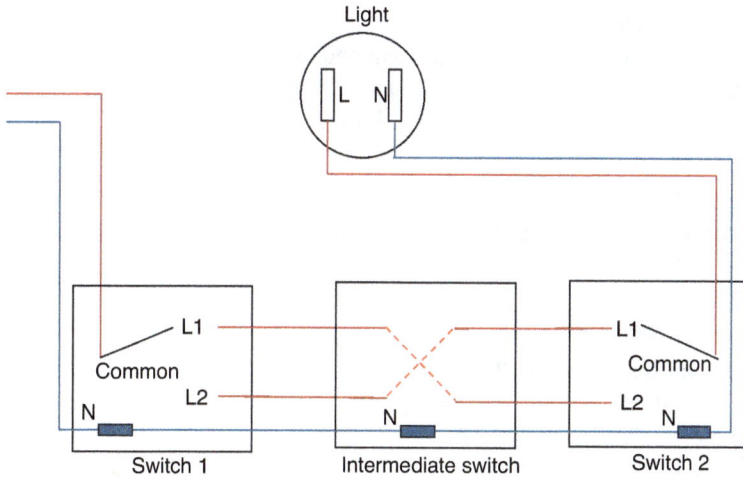

Figure 15.8   Two-way-and-intermediate switch wired in singles (earth omitted for clarity)

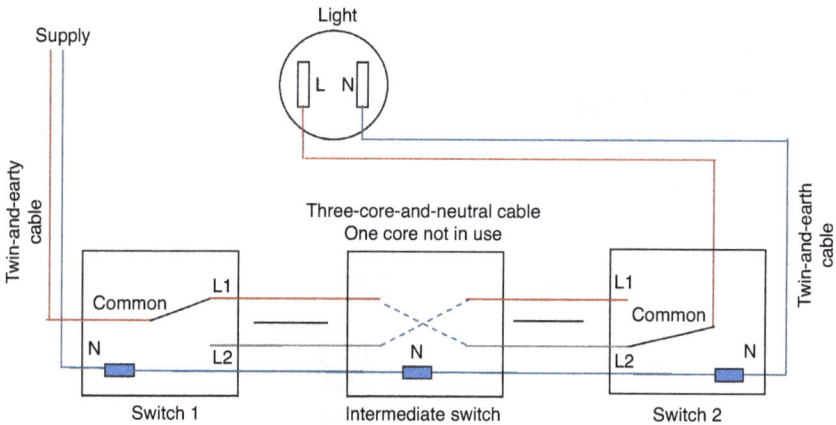

Figure 15.9   Two-way-and-intermediate switch wired in multi-core cables (earth omitted for clarity). The spare core of the three-core cable could also be used as the linking neutral between the two switches, provided it is clearly identified as a neutral with blue sleeving.

Loop-in method

Many existing installations where the lights are wired in twin-and-earth and three-core-and-neutral cables will be connected differently to the diagrams shown here. Instead of taking the line feed and neutral straight to the switch, they are terminated at the light itself.

1. There is an extra set of terminals in most ceiling roses. These are marked as 'loop-in' terminals.
2. The line feed is terminated here with the brown conductor from the switch twin-and-earth.
3. The blue conductor in the switch twin-and-earth is used as the switch wire and is connected into the line terminal in the ceiling rose.
4. Because the blue conductor is used as a line instead of neutral, it must be marked as such with brown sleeving at both the switch and light terminals.

The reason that the loop-in method has been largely superseded is because modern lights, for example, ELV lights with transformers do not have a loop-in facility (Figure 15.10).

Light-emitting diodes

The light-emitting diode (LED) has revolutionised lighting. This type of light can be used for small, battery-operated novelty lights and can be manufactured in flexible strips, some of which can be cut at any point with scissors and still work. Uses for LED lighting include car headlamps and domestic and industrial lighting.

LEDs are ideal for digital lighting systems. They work using semiconductors with their own solid-state circuit boards, which means that they are responsive to direct digital signals. They can also produce dynamic lighting effects and colours.

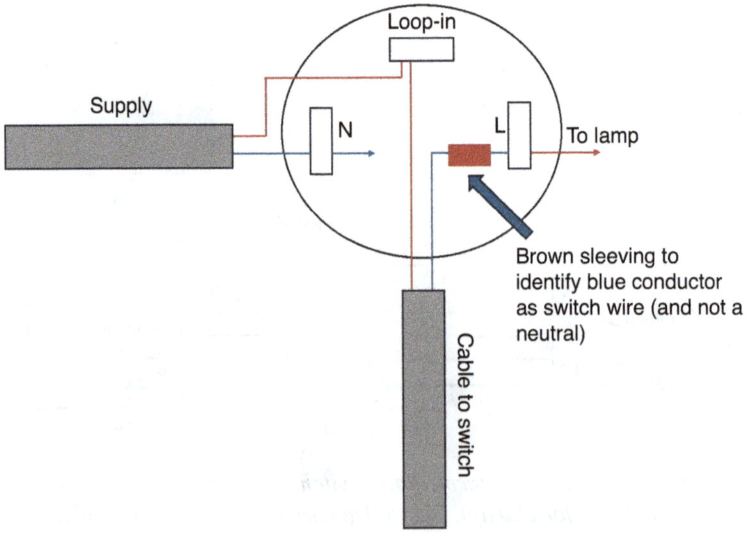

Figure 15.10   Loop-in method for wiring lights using twin-and-earth cable (earth omitted for clarity)

The advantages of this type of lighting are:

- Energy saving – the control system can be programmed to respond to light levels and only energise the lights when light falls below a certain level.
- Actual light levels can be closely automatically controlled.
- Each light can be individually controlled.
- Extra low voltage only is required.
- There are a variety of control methods, e.g.:
  - ○ Pre-programmed
  - ○ Light sensors
  - ○ Voice activated

Modular lighting systems

These are versatile lighting systems that do not rely on fixed wiring. Instead, a track is installed, fed by a single supply which means that the lights themselves can be plugged in then easily changed without disconnection and re-connection of wiring. The tracks themselves are essentially a mini-busbar system. Switching can be manual or wireless, which further reduces the need for wiring. The busbar voltage can be transformed down to extra-low voltage to accommodate LED lighting.

Digital lighting systems

Digital lighting systems are, essentially, systems where lighting can be controlled automatically or remotely using a digital signal rather than by mechanical switches. Building management systems (BMS) often include lighting within their control scope. In domestic installations, digital lighting can be operated via voice activation using a smart device (Figure 15.11).

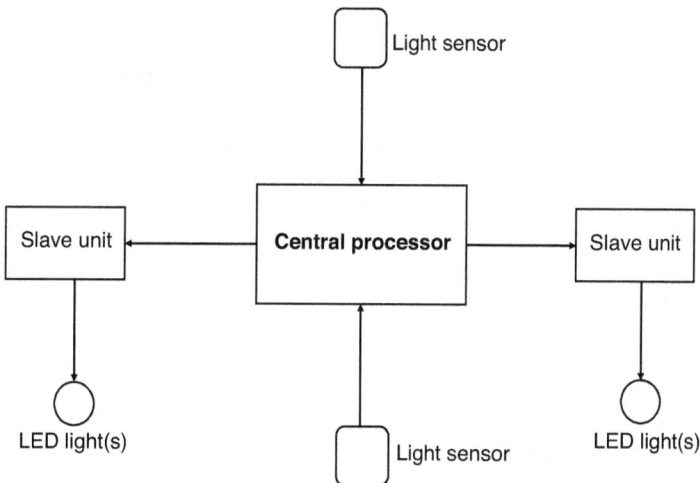

Figure 15.11 Simplified diagram of a digital lighting system. A central processor receives commands from sensors or other inputs and then passes on commands to the appropriate slave units that, in turn, operate the lights themselves.

Kinetic/quinetic switching

A kinetic/quinetic switch is one that operates a light using a wireless signal rather than via a hard-wired connection. The word *quinetic* is a derivation of kinetic. Kinetic energy is created by movement. The action of pressing or swiping your finger over the switch will generate the required energy using a micro energy acquisition module. This is, in effect, a tiny generator that produces just enough power for a wireless signal. The signal it creates is sent from the switch to a controller that energises the appropriate light(s).

The advantage of these types of switches is that they can be placed anywhere in a room and easily moved to an alternative location. Quinetic switches can also be used for dimmable lamps.

15.5   13 A socket outlets

The IET On-Site Guide and BS 7671 both give three configurations for wiring 13 A socket outlets.

- A1: Ring final circuit – 32 A protective device, 2.5 mm^2 cables.
- A2: Radial – 32 A protective device, 4.0 mm^2 cables.
- A3: Radial – 20 A protective device, 2.5 mm^2 cables.

We have already discussed radial circuits, so we will concentrate on the ring final circuit for this chapter.

Unlike most loads, a socket outlet is a *means* of connection. In any given circuit, some of the socket outlets may remain unused. Also, there is no set load, because a number of different appliances can be plugged into the circuit. The maximum load that can be connected via a 13 A plug is 3 kW. This, combined with the fact that the number of 13 A socket outlets supplied by a ring final circuit within a 100 m^2 area is unlimited, means that a socket outlet circuit could, potentially, feed a heavy current load or be lightly loaded.

Each load plugged into the ring final circuit must be protected by its own fuse, usually in the plug top.

It can be seen from Figure 15.12 that, unlike a radial circuit, when a single-line circuit takes current to the load and a single neutral conductor returns the current from the load, a ring final circuit consists of two line, neutral and protective conductors. In effect, a line conductor is wired to each 13 A socket outlet and then returned to the consumer unit from the last one. The same applies to the neutral and to the CPC.

The ring final circuit is, therefore, fed from each end. This method is adopted for two reasons.

Firstly, it limits the distance travelled by the current. The middle socket outlet is now the furthest point rather than the last socket in a radial string.

A ring final circuit is, in effect two circuits feeding one load. These circuits are wired in parallel (and are called *legs*). We have already seen in chapter 4 that the equivalent resistance (Re) for two identical resistances in parallel can be calculated

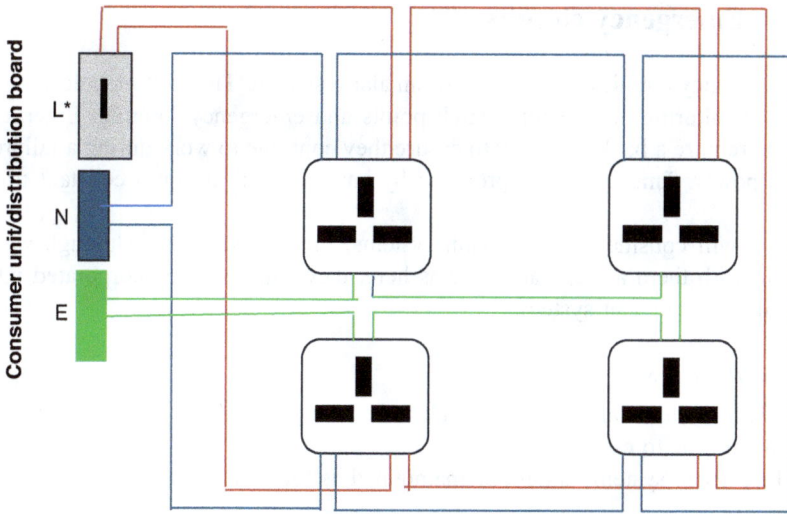

*Line conductors connected into protective device

Figure 15.12   A1 Ring final circuit. Note that each conductor returns to the source after connection of the last socket outlet.

as follows:

$$R_e = \frac{\text{Resistance of one load}}{\text{Number of resistances}}$$

Therefore, for a ring final circuit:

$$R_e = \frac{\text{Resistance of one leg of ring}}{2}$$

Because the equivalent resistance is half of the resistance of one 2.5 mm^2 cable, it will carry twice its normal current. Rule-of-thumb states that a 2.5 mm^2 conductor will carry a maximum of 16 A. Because its resistance is halved, it will carry 32 A. The Design Chapter in Volume 2 will prove this in more detail.

Spurs

Extra 13 A socket outlets or feeds to other loads can be wired off the ring final circuit as spurs. These are radial circuits connected either to one of the 13 A socket outlets in the ring, or from a junction box. If the spur feeds:

- 13 A socket outlet(s) it can feed one single or one two-gang socket outlet
- A directly connected load such as a heater must be connected to the load via a switched fused isolator that provides close protection to that load.

The number of spurs must not exceed the number of 13 A sockets in the ring final circuit.

15.6   Emergency circuits

An emergency circuit is one that feeds an alarm system. This will include a control panel, monitoring equipment, switch points and emergency lighting. Emergency circuits require a backup supply to ensure they continue to work during a failure of mains power. This is usually provided by batteries that are on a constant trickle charge.

We will consider the two main systems, fire and security. Although we are dealing with them as separate systems here, they can also be incorporated into a building management system.

Fire alarm system

Fire alarm systems are designed to alert the occupants of a building to a fire and give them time to escape.

Fire alarm systems should be maintained as follows:

- A weekly test (carried out at the same time each week)
- Operation of a manual call point every week (tested in rotation)
- Monthly test to simulate loss of power (if an emergency generator is used)
- Recording information in a logbook

These are some of the main types of fire alarm system.

Analogue

For an analogue system, the premises is divided into zones. Each zone contains all the components required for fire monitoring and alarms. When a fire occurs, the alarm is triggered. The central main panel then indicates the zone in which the fire has broken out.

Note the resistors at the end of each Zone circuit in Figure 15.13. The purpose of these is twofold:

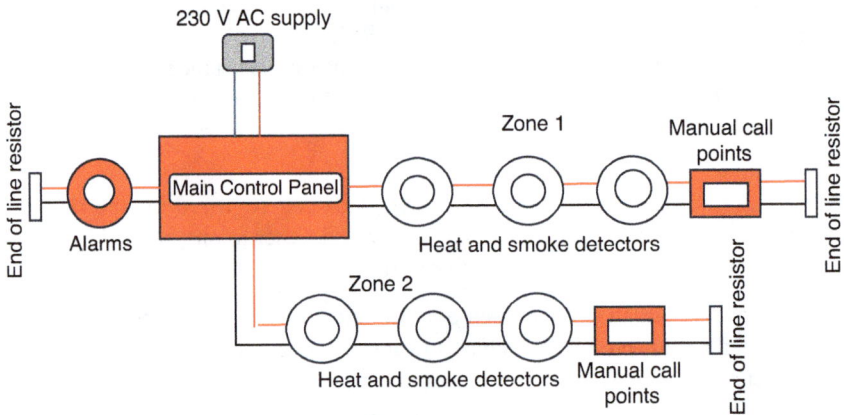

Figure 15.13   Analogue fire alarm system

- To close the circuit and limit current.
- As a maintenance alert – if the resistance increases, this means that there is an open circuit that is indicated by the control panel with a message such as 'Open circuit – Zone 1'.

Addressable

Instead of dividing the premises into zones, an addressable system is able to pin-point the exact location of a fire. This can save time and possibly lives by directing fire and rescue services to where they are needed.

Because it is digital, an addressable system is easier to integrate into a building management system (see Section 15.7). When fire is detected, the BMS can not only activate alarms and **fire suppression**, but it can also bring lifts to a suitable floor, close fire doors, open escape exits and alert the emergency services (Figure 15.14).

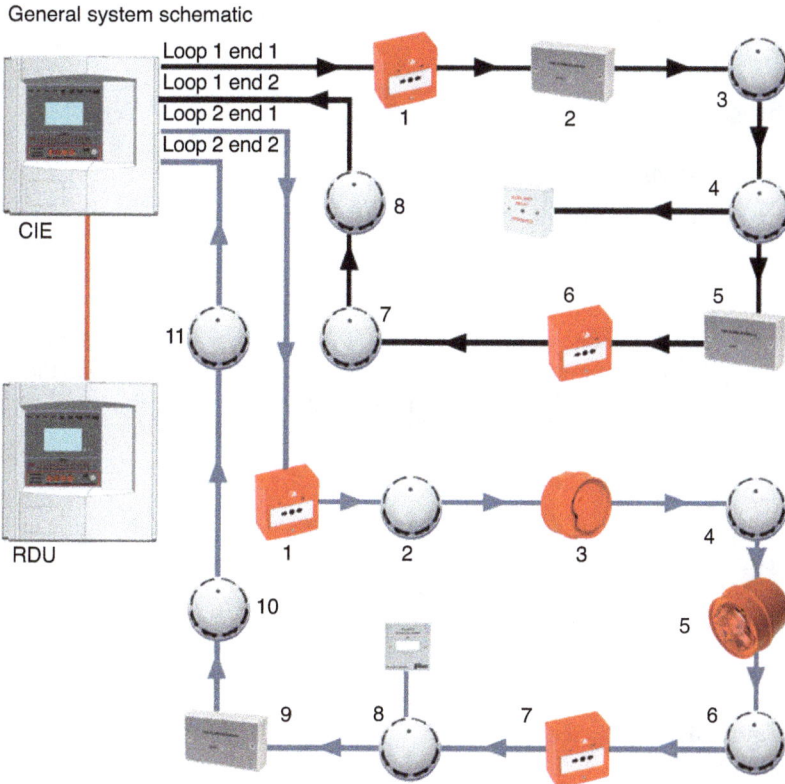

Figure 15.14 Addressable fire alarm system (by kind permission of Fike Corporation)

KEY TERM: Fire Suppression – Means by which a fire can be extinguished or contained automatically, for example, water sprinklers or injection inert gas into the affected area.

Linear heat detection

An ideal system for tunnels and large spaces such as factory floors and warehouses. A cable with zinc-coated steel conductors insulated by polymer-coated thermoplastic is run through the area to be protected. When a fire breaks out, the insulation melts. The resulting short circuit between the two conductors, and its exact location, are detected by the system's controller, and an alarm is activated.

An alternative version is digital temperature sensing. In this case laser pulses are sent down a fibre optic cable. These pulses monitor temperature and alert the control panel when there is an increase above a predetermined level. Uses for digital temperature sensing:

- Tunnels, particularly those that are hard to access
- Dusty environments where conventional heat and smoke sensing is hampered by the dirty atmosphere
- Conveyor belts
- Battery racks

Components of a fire alarm system

The main components of an emergency system are as follows.

Mains supply

While the power throughout the system may be low-voltage DC, the main panel and some of the components will require a 230 V main supply. For a control panel, this supply is usually terminated in a switched isolator, then connected to the panel via a flexible cable. The general wiring must be in fire-resistant cable such as FP or mineral insulated.

Control panel

The control panel acts as a central processor for the system. It receives signals from call points and monitors such as heat, smoke or gas detectors and responds accordingly.

Sensors

These are the eyes and ears of the system. These could be:

- Smoke detector
- Heat detector

Emergency call points

As well as automatic heat and smoke detection, all commercial, public and industrial premises must be equipped with manual call points to be operated by anyone who discovers a fire. This is usually in the form of a break-glass call point. The glass front holds in a normally closed switch. When the glass is broken, using the hammer provided with the device, the switch is released and the alarms are energised.

Alarms

Alarms are sounded in the event of a fire to warn the building's occupants and trigger an evacuation. These can be in the form of a bell or a siren.

Emergency lighting

The purpose of emergency lighting is to provide illumination for evacuation of a building if the main power fails. Emergency lights can be individual, specialised lights or a selected number of the existing luminaires. The backup batteries necessary for emergency lighting are generally lead-calcium.

The lights themselves can be:

- Halogen with xenon filaments
- LEDs either overhead or as pathway strips on the floor that guide people to fire exits.

Maintained

Always on during both normal and emergency conditions. They are supplied by the mains supply but are equipped with backup batteries to ensure operation during a power failure.

Non-maintained

Only come on during emergency conditions. They are illuminated by a battery pack, which is on constant charge from a mains supply.

Intruder alarms

To protect against trespass and theft, intruder alarms use a number of components to maintain the security of a premises. They employ both:

Detection

To discover the presence of intruders and issue an alarm and alerts, including a direct alert to the emergency services.

Deterrent

To discourage would-be intruders using lights and alarms.

Components of an intruder alarm system

Control panel

As with all emergency and monitoring systems, the core of an intruder alarm system is a control panel that receives information from the outlying sensors and alarm points and then sends out appropriate instructions to alarms, etc. The premises will be divided into zones to localise the security breach.

Sensors

This can be a number of different types:

- Door and window sensors – switches located in doors and windows that operate if they are opened.
- Movement detection – PIR sensors and light that use an infrared beam. If the beam is intact, the sensors read this as normal conditions. If the beam is broken, then a light or alarm is operated.
- Heat and vibration sensors

Cameras

Cameras can either be in constant use or triggered by the presence of an intruder. They can be used to identify intruders. They can also act as a deterrent, their presence being obvious and a warning to keep away.

A multiplexer processes information from multiple cameras and adds a code mark to each image. The images can then be stored on a single drive. Any image can be recalled using the code mark.

Alarms

Bells or sirens can be used to both sound an intruder alert to the occupants or owners of a premises, and also as a deterrent to the intruders themselves. These can be local but also connected directly through to the emergency services.

15.7   Building management systems (BMS)

The **building engineering service** components in modern commercial, public and industrial buildings are generally controlled centrally by a building management system (BMS).

> **KEY TERM: Building engineering services – The systems that make a building work, for example:**
> - **Electrical installation**
> - **Plumbing**
> - **Heating and ventilation**
>   **Could be described as the arteries and nerves of a building.**

There is no limit to the services that can be controlled by a BMS. Among the most common are:

- Lighting
- Air conditioning
- Humidity
- Heating
- Alarms
- Security

- Backup power
- Opening and locking doors

They can also be programmed to operate blinds and windows.

At the heart of a BMS is a central computer system, or processor, into which the settings for each function are programmed. The central processor constantly communicates with the various sensors located throughout the building via a series of outstations. These sensors monitor heat, lighting, humidity, etc.

The processor interrogates its own settings to ensure that the levels are within acceptable parameters. If not, it instructs the necessary changes to be put into effect by the outstation. The outstations are linked and constantly share information. The point at which the changes to the environment and so forth are adjusted are called actuators. This could mean a valve or switch or any other device that operates an element of a building engineering system.

Because a BMS exercises close control over a building's engineering and environmental systems, it is able to reduce energy consumption and cost. They can also be programmed to carry out automated maintenance and alert building services staff to problems and faults.

Three features of a BMS are (Figure 15.15):

- Communication **protocol** – A system of digital messages, rules and formats that are used to exchange messages between the elements of the system.
- BACnet – Building Automation and Control is a protocol designed to integrate differing building control protocols arising from the various components in the BMS.
- LON – Local Operation Network, which is a network used to link the separate elements of a BMS and enable them to exchange information.

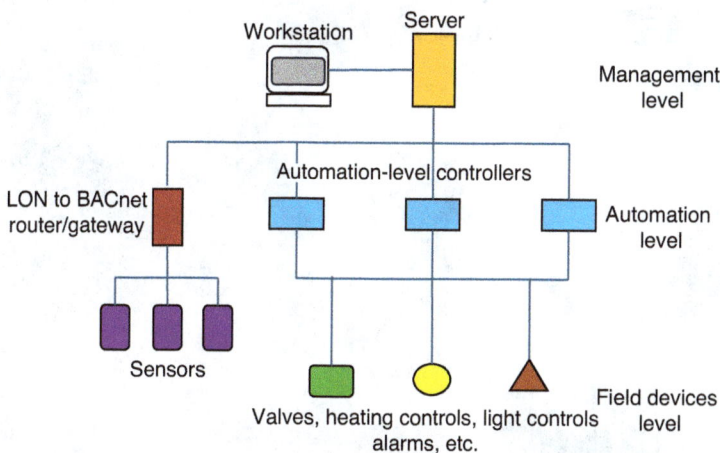

Figure 15.15 Basic components of a BMS system

KEY TERM: Protocol – A method by which computers and other devices can communicate even if they run on different software and processes.

15.8   Standby supplies

For many, loss of main electrical supply is an inconvenience but no threat; however for more critical areas, such as hospitals, emergency services and some industrial operations, power shutdowns can be disastrous, both in financial and human terms. So, it is essential that a standby power system is available to provide the necessary supply for critical areas until mains power can be restored. These systems are usually brought online automatically the moment the supply fails.

Backup supply hazards

The main risk posed by standby supplies is that when the power is isolated for maintenance or repair work, the standby system might be brought online. An installation, believed to be isolated and safe, will be re-energised, with the associated shock hazard. It is vital that before any major work takes place on an existing installation, it should be ascertained that there are no alternative power supplies feeding into the installation. This includes renewable energy systems such as photovoltaic arrays (solar panels – see Volume 2).

Standby generators

The mainstay of backup power is the generator (Figure 15.16). Critical areas will be connected to both the mains supply and the generator. The generator will be started

Figure 15.16   A standby generator. Image credit: Mikael Häggström, via Wikimedia Commons, CC0, Public Domain.

the moment mains power fails. These generators are physically large and powerful enough to sustain the required current load.

BS7671 states that supplies provided by a generator must have the same electrical protection as the mains supply, that is, against shock and overcurrent. It also states that it must not come online while the mains supply is working. The following precautions must be taken (quoted directly from BS7671):

1. An electrical, mechanical or electromechanical interlock must be in place between the operating mechanisms or control circuits of the changeover switching devices
2. A system of locks with a single transferable key
3. A three-position break-before-make changeover switch
4. An automatic changeover switching device with a suitable interlock
5. Other means providing equivalent security of operation

The components of a generator are as follows:

Engine

Usually powered by natural gas or propane, although some are diesel-fuelled, the engine provides the mechanical energy to drive the electrical generator head. These engines are designed to run for long periods of time without excessive wear and are supplied directly with fuel. They have an auto-start function. They will need regular maintenance to ensure that they will start when needed.

Generator head

Sometimes called the alternator, this is the AC machine that produces the electrical energy. A rotating magnetic field induces current into a set of stationary, or stator, windings. The output is AC, produced at the same frequency as the installation.

Voltage regulator

To ensure a regular and stable supply, the regulator monitors the output voltage and adjusts it if necessary. This is mainly to protect any sensitive electronic circuits.

Start time

When the mains power failure is detected by the generator control system, the machine is started automatically. However, suitable voltage production is not instantaneous. Modern generators can be online in less than 10 seconds but it can also be as long as 30 seconds.

Uninterrupted power supply (UPS)

In certain cases, such as an intensive care unit or a critical computer system, any delay between loss of mains power and standby generator operation can be catastrophic. Uninterrupted Power Supplies (UPS) are designed to bridge the gap between power loss and the generator coming online. BS7671 states that a UPS must be capable of operating the circuits it supplies and also able to allow the operation of their protective devices.

As the name suggests, a UPS will immediately begin feeding critical circuits the moment the supply is lost. The main components of a UPS are as follows.

Batteries

A bank of batteries forms the main part of a UPS. These produce the required amount of power for the required time. They will discharge relatively quickly but are not required to provide standby power for a long period.

Inverter

Converts the DC produced by the batteries to AC. The inverter also synchronises the UPS output with the mains supply frequency.

Rectifier

A rectifier converts AC to DC as part of the constant, or trickle, charging system for the batteries.

Static bypass switch

An electronic device that switches between states. When the mains power fails, it brings the UPS online and takes it offline when either the generator takes over or the mains supply is restored.

15.9   Exercises

1. Give the three reasons BS7671 gives for dividing an electrical installation into separate circuits.
2. Describe how the sub-circuit neutrals should be connected in a board with its ways divided between a number of RCDs
3. What is good practice when connecting the CPCs and neutrals in a consumer unit?
4. What is the main difference between a functional switch and an isolator?
5. Draw a circuit diagram for a two-way-and-intermediate switching arrangement, wired in multicore cable. Include two intermediate switches and two lights in the diagram. You can omit the CPC for clarity.
6. What is the main difference between a ring final circuit and a radial circuit?
7. Give two reasons for wiring 13A socket outlets as a ring final circuit.
8. List the main components of, and their functions in, an analogue fire alarm system.
9. What are the two main types of intruder alarms?
10. What is meant by building engineering systems and list two examples.
11. What is a building management system?
12. Why is an uninterrupted power supply necessary when a standby generator is already in place?

15.10 Multiple-choice questions

15M1 A double-pole switch breaks and makes:
(a) Line and neutral
(b) Line only
(c) Line, neutral and earth
(d) Neutral and earth

15M2 How many live conductors are there in i) a single-phase supply ii) a three-phase and neutral supply?
(a) i) Two ii) Three
(b) i) One ii) Three
(c) i) Two iii) Four
(d) i) One ii) Four

15M3 Isolation of an electrical supply so that a bench saw blade can be changed is:
(a) Replacement
(b) Mechanical
(c) Industrial
(d) Electrical

15M4 Isolation of an electrical supply so that extra 13A socket outlets can be added into an existing ring final circuit is:
(a) Replacement
(b) Mechanical
(c) Industrial
(d) Electrical

15M5 How should an isolator be rated?
(a) Full load current
(b) No current
(c) Half-load current
(d) Instantaneous load current

15M6 Which of these is NOT a functional switch?
(a) Light switch
(b) Contactor
(c) Hob unit on/off switch
(d) 13A plug and socket outlet

15M7 How should a light switch be rated if it switches discharge lighting?
(a) $2 \times$ circuit design current
(b) $1.8 \times$ circuit design current
(c) $0.8 \times$ circuit design current
(d) $1.8 \times$ protective device rating

15M8   How many conductors are there in an A1 13A ring final circuit?
(a) 6
(b) 3
(c) 5
(d) 9

15M9   What is the protective device rating and minimum conductor size for an A2
radial circuit feeding 13A socket outlets?
(a) 20 A/2.5 mm^2
(b) 32 A/2.5 mm^2
(c) 32 A/4.0 mm^2
(d) 20 A/4.0 mm^2

15M10 What is the maximum number of spurs allowed for a ring final circuit with
12 × 13A socket outlets?
(a) 6
(b) 1
(c) 12
(d) 24

15M11 What are the two main types of fire alarm system currently available
(a) Answerable and anti-nuisance operation
(b) Anti-lock and assisted
(c) Anti-lock and anti-nuisance operation
(d) Addressable and analogue

15M12 Why are resistors fitted to the ends of smoke and heat detection circuits?
(a) Limit signal and maintenance alert
(b) Limit current and maintenance alert
(c) Limit inductance and overcurrent alert
(d) Limit resistance and overcurrent alert

15M13 What is the heart of a BMS system?
(a) Sensors
(b) Command protocol
(c) Central processing unit
(d) Voltage regulation protocol

15M14 What component ensures that a standby generator supply is stable?
(a) Current regulator
(b) Power factor limiter
(c) Power limiter
(d) Voltage regulator

15M15 What is the purpose of an inverter?
(a) Converts AC to DC
(b) Changes AC frequency
(c) Converts DC to AC
(d) Changes DC pulse magnitude

Chapter 16

Wiring systems

So far, we have looked at the principles of science that underpin electrical installation and engineering. We have also considered the supply and the installation
itself, in terms of protection and types of sub-circuit. This chapter describes the
methods by which these circuits are actually wired. It is important that the correct
wiring system is selected for an installation. Some offer only the most basic
mechanical protection to cables, and others provide a high level suitable for harsh
environments such as factories and farms.

The method of installation is also relevant to the size of a cable. Cables run
inside containment such as trunking and conduit will be warmer than cables clipped
directly to a surface or on a cable tray. As temperature rises, so does resistance, so
this may mean that larger cables are required to feed a particular load.

16.1   Conductor materials and construction

Cable conductor manufacture

A cable conductor is cold drawn, which means that it is reduced to the required size
using rollers and dies.

Cold-drawn conductors are too rigid and hard for practical use, so they are
annealed by heat treatment, which causes them to become comparatively soft and
pliable. Continuous bending of the conductors, however, particularly if the strands
are large, will result in them being work-hardened and increases the risk of fracture.

Basic electrical conductors have already been considered in Chapter 3, showing that from a purely electrical point of view, silver is the best conductor, but that
its poor mechanical properties and high cost rule it out as a cable conductor.

Copper

Copper is second only to silver as an electrical conductor. It is, however, a much
more practical proposition because it is cheaper and is easily cold-drawn to make
wires. All small cable conductors are made of copper.

Aluminium

Although aluminium is more than half as resistive again as copper, its density is
less than one-third. This means that an aluminium conductor, although larger, will
be lighter than an equivalent copper conductor. The disadvantages are:

- lower strength
- higher thermal expansion
- rapid oxidation that necessitates special jointing techniques

Aluminium conductors are almost always used for heavy-current overhead lines (often with a steel core for extra strength) and have been widely used for power cables. BS 7671 does not allow the use of aluminium cables of a cross-sectional area less than 16 mm². Table 16.1 shows the relationship between copper conductors and equivalent aluminium ones.

Solid and stranded conductors

Some small cables (≤ 2.5 mm²), mineral-insulated and some aluminium power cables have single-strand or solid conductors. The majority of cables, however, have stranded conductors to make them more flexible for installation purposes. The standard arrangements are for 1, 7, 19 or 37 strands. The strands in flexible cables are more numerous but smaller strands (Table 16.2).

It is possible to obtain standard wiring cables up to, and including, 2.5 mm² with either stranded or solid conductors. Stranded versions are used in situations where their extra flexibility is useful.

Table 16.1   Standard sizes of copper cables with comparable aluminium sizes

Copper-conductor CSA (mm²)	1.0	1.5	2.5	4	6	10	16	25	35
Aluminium conductor (of equivalent current rating) CSA (mm²)	1.5*	2.5*	4*	6*	10*	16	25	30	50

*The use of these cables does not comply with the IEE Wiring Regulations (BS 7671).

Table 16.2   Stranding of copper cable

Cross-sectional area (mm²)	The first number indicates the number of strands; the second number gives the diameter of each strand in mm	
	Cables	Flexibles
0.5	–	16/0.20
0.75	–	24/0.20
1.0	1/1.13	32/0.20
1.5	1/1.38	30/0.25
2.5	1/1.78	50/0.25
4	7/0.85	56/0.30
6	7/1.04	84/0.30
10	7/1.35	80/0.40
16	7/1.70	126/0.40
25	7/2.14	196/0.40
35	19/1.53	276/0.40

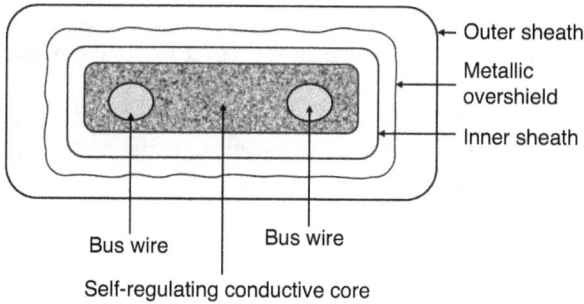

Figure 16.1   Heating or trace-heating cable

Heating (or trace) cable

Most electrical conductors are constructed so that they have low resistance. This results in small power losses and low temperature rises within the cable itself. Sometimes, however, cables are used for heating, and their conductors must have a higher resistance. Such conductors are usually alloys containing some combination of copper, iron, steel, nickel and chromium (Figure 16.1). Examples of heating cable use are:

- underfloor heating
- keeping roads and ramps clear of snow
- keeping gutters free of ice
- soil heating for sports and horticultural purposes

There are different types of heating cables.

Constant wattage heating cables

These provide consistent heat along their length. They are unaffected by changes in **ambient** temperature and are commonly used where steady heat is required.

> **KEY TERM: Ambient Temperature – The temperature of the surrounding environment, for example, the temperature in a room.**

Self-regulating heating cables

These cables are able to change their power output as the ambient temperature rises and falls. The cable heat output will increase when the temperature drops and decrease when it rises. As a result, they are energy-efficient and ideal for environments where frequent temperature fluctuations occur.

These cables work through the relationship between temperature and resistance.

- When the temperature drops – the resistance of the self-regulating core is reduced, allowing higher current to flow between the bus wires' cores. Increased current flow means higher power production.
- When the temperature rises – the resistance of the self-regulating core is increased, allowing less current to flow between the bus wires' cores. Reduced current flow means lower power production.

Series resistance heating cables

These cables are designed to provide a uniform heat for long runs such as industrial pipework and other specialised applications. The core of a series resistance, or trace heating, cable has a high resistance so that heat is produced when current flows through it. For pipe runs of several kilometres, extra power supplies may be required at intervals along the run.

16.2   Cable insulators

Plastic

Polyvinyl chloride (PVC) (thermoplastic)

A basic plastic material that can be made in many forms. BS 7671 refers to PVC as a 'thermoplastic' material. This means that its form can be changed if its temperature exceeds a certain level. PVC is:

- Robust
- Chemically inert
- Has good ageing and fire-resisting properties

It is more resistant to weathering and sunlight than rubber, but will harden and crack in the presence of oil or grease. High temperatures lead to softening and possible insulation failure, whereas at low temperatures, PVC may become brittle. Very little absorption of water will occur, although PVC is inferior to polychloroprene in this respect. PVC is widely in use as a sheathing material as well as for insulation. It can only usually be used where conductor temperature does not exceed 70 °C. The material emits dense smoke and corrosive fumes when it burns.

Polychloroprene (PCP) (thermoplastic)

Polychloroprene, also known as Neoprene, is a plastic material of lower strength and lower insulation resistance than PVC. However, it is more resistant than PVC to weathering and to attack by:

- Oils
- Acids
- Solvents
- Alkalis
- Water.

PCP is more elastic than PVC, and this elasticity is not affected by increased temperatures. Although more expensive than PVC, PCP is used as a cable insulation and sheathing material for conditions where PVC would not be suitable.

Chlorosulphonated polyethylene (CSP) (thermoplastic)

This material is mainly used for sheathing cables insulated with other plastics. It is capable of operating over the temperature range −30 °C to 85 °C, and has excellent oil resisting and flame-retarding properties. It is very tough, and very resistant to heat and water. This is the sheathing material that is called 'HOFR' (heat resisting, oil resisting and flame retarding) in BS 7671.

Cross-linked polyethylene (XLPE) (thermoplastic)

This is a polymeric insulation (like EPR above) and is used particularly for power cables (Figure 16.2). In this application it shows cost advantages over oil-filled and pressurised cable insulation, not least in the simpler jointing techniques that may be used. The material does not burn easily, and when it does, only limited emission of smoke and fumes occurs. Used for the outer sheath and insulation on steel wire armoured cables in situations where low smoke emission is important (Figure 16.3).

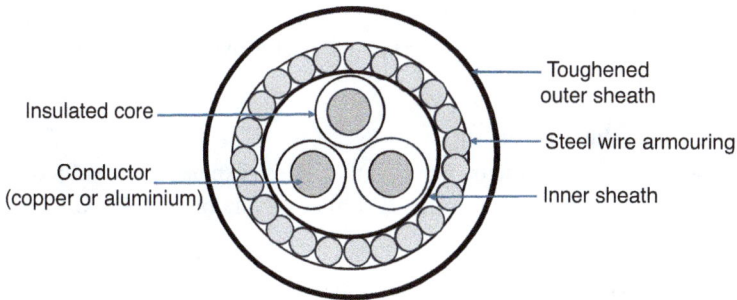

Figure 16.2   Cross-section of typical steel-wire armoured cable

Figure 16.3   Steel-wire armoured cables. Image credit: Pawarbhushan, via Wikimedia Commons, CC BY-SA 4.0.

Rubber

Vulcanised rubber (VR) (thermosetting)

A good electrical insulator with high flexibility. BS 7671 now refers to VR as a 'thermosetting' material, meaning that its form cannot be changed after manufacture. It is subject to rapid ageing and cracking due to weathering and exposure to sunlight, and softens and becomes sticky in the presence of oils and greases. It ages rapidly at high or low temperatures. It is not widely used but is available in a form that will operate safely up to 85 °C.

Silicone rubber (thermosetting)

Silicone rubber is a synthetic material with many of the advantages of natural rubber. It has good weathering properties and will resist attack from water and mineral oils, but not from petrol. It remains elastic over extremes of temperature (−70 °C to 150 °C).

Butyl rubber (thermosetting)

Butyl rubber is another synthetic material that is less expensive than silicone rubber but has similar advantages. It remains flexible over the range −40 °C to 85 °C, but will burn readily once ignited.

Ethylene–propylene rubber (EPR) (thermosetting)

EP rubber is generally similar to butyl rubber but with improved properties. It is resistant to heat, water, oil and sunlight, and is suitable for direct burial, exposure to weather and contaminated atmospheres. It is used to insulate power cables that have an increased current rating owing to its heat resistance.

Other cable insulation types

Magnesium oxide

This material is in the form of a white powder and is invariably used in mineral-insulated cables. It also finds application in some heating elements. Magnesium oxide is non-ageing and will not burn but is hygroscopic (absorbs moisture from the air), when it loses its insulating properties. It is unaffected by high temperatures and is a good conductor of heat.

Paper

Dry or oil-impregnated paper, wound in long strips over the conductors, was once the most common method of insulating underground cables but has now been overtaken by XLPE. As long as the paper is kept dry (usually by sheathing with lead alloy), its insulating properties are excellent.

Paper is sometimes used to insulate transformer windings.

Glass fibre

Glass fibre, impregnated with high-temperature varnish, is used to insulate some flexible cords used for high-temperature heating and lighting applications.

16.3   Cable types

Power cable selection

Cables are selected for two main reasons:

- The type of installation – we have already seen that different cables are more suited for certain environments; for example, a factory would require hard-wearing cables with good levels of mechanical protection.
- Current-carrying capacity – the cross-sectional area of a cable depends on the amount of current it needs to carry. In other words, the cable needs to be large enough to carry the required current. There are several other factors involved in the process of selecting the correct-sized cable for a circuit, which are discussed in the Design Chapter in Volume 2.

Single-core

The simplest and most basic form of cable is the single core. As the name suggests, the cable boasts only one conductor so in a single-phase-and-earth supply, for example, three separate cables are required.

Single-core is manufactured both in sheathed and unsheathed versions. An example of sheathed single-core is the meter tails that connect the DNO electricity meter to the consumer unit. Unsheathed single-core is used where containment such as conduit or trunking is required. Because the containment provides mechanical protection, there is no need for the cable to be sheathed. In fact, unsheathed cables *must be* installed in containment. No part of a circuit wired in single-core unsheathed cables should be exposed to touch.

There are other single-core cables in use, for example, as high-voltage supply cables and single-core armoured types.

Twin-and-earth

BS 7671 calls these cables flat profile, multicore cables, which is probably a more accurate name for them because four- and three-core versions are also available. However, this type is generally known as twin-and-earth, which is why they are given that name here.

The cable consists of:

- An outer layer of PVC, formed into a flat oval shape.
- The insulated and colour-coded line and neutral cores
- An uninsulated circuit protective conductor (CPC) – the earth protection conductor. This conductor is left bare because it does not carry current under normal working conditions. As a result, the cable is lighter and more flexible. However, the CPC must be covered by a green-and-yellow sleeving wherever it is exposed, such as at the connection point of an electrical accessory.

Twin-and-earth cables are versatile and hard-wearing. While not suitable for an industrial or harsh weather environment, they are ideal for domestic and office-type installations.

Sheathed cables can be:

1. Fixed directly to a surface using cable clips
2. Laid freely across roof spaces
3. Run through containment
4. Buried in plaster walls
   (a) ≤50 mm depth with added protection of a conduit or capping plus 30 mA RCD at circuit source
   (b) 50 mm with added protection of a conduit or capping
   (c) If run is horizontal, then cable must be run in buried metal conduit

5. Run under floorboards – the cable must pass through holes that are at least 50 mm below the top of the joist to prevent damage by nails. The IET On-Site Guide shows the maximum dimensions of any joist holes or slots required.

Sheathed cables should not be manipulated or installed when the temperature is below 5 °C, or there is a danger that the sheath or insulation may crack. Neither should they be exposed to strong ultraviolet, such as sunlight. Twin-and-earth cables were once considered suitable for outdoors wiring but once the effects of UV were understood, they were no longer recommended for this type of work.

If they are tightly bent, the sheath may fracture; the minimum bending radius should never be less than three times the overall cable dimension and is larger for cables with conductors >10 mm^2.

Steel-wire armoured (SWA)

Another versatile cable, SWA is an ideal choice for installations in harsh environments. They can be used as circuit wiring but also for supplies. Uses for SWA include:

- Feeds to industrial loads such as machines
- Supply cables
- Sub-main cables distributing power throughout a large premises
- Underground cables

Construction of a typical SWA cable.

- Toughened PVC outer sheath
- Steel wire armour, spiralled along the cable's length
- PVC inner sheath
- XPLE-insulated cores
- Conductors that can be
  ○ Copper
  ○ Aluminium
  ○ Solid; circular or wedge-shaped
  ○ Stranded; circular or wedge-shaped

A common installation method for SWA is to run it on a tray or ladder. It can also be clipped directly to the surface using cleats. When installed underground, SWA can be run through ducts but is robust enough to be buried directly. A warning tape that indicates the presence of a buried cable must be laid 150 mm above the cable itself.

SWA termination

A specialised gland is required for termination of an SWA cable (Figure 16.4). The armouring can be used as a CPC, so it needs a firm connection to the gland and thence to the installation's earthed metalwork. An SWA should enter an accessory, consumer unit or distribution board through a suitably sized hole, where it is terminated into a gland and anchored in place.

The gland should include a banjo-type earth connecting ring. This is clamped between the locknut in the termination enclosure and the metalwork. An earth conductor is connected to the tab on the banjo and the enclosure's earth terminal. This is particularly important if the armouring is used as the CPC.

Flexible cable

Flexible cable, or flex (the term used from now on in this section), is generally used to connect an item of equipment to a **fixed wiring** supply, for example, a plug and socket or a switched-flex outlet point. There are numerous types of flex, so only the main types will be described here.

BS7671 allows for very few installations to be wired in flex. Exceptions include caravans, in which the wiring *must* be comprised of flex. This is to take into account the twisting and movement that occurs in a caravan when it is being towed and during normal use.

Flex doesn't need to be fixed, and short runs can be looped through air from supply to load. When used as an extension lead, a flex must be a three-core type. Some loads may not require an earth but others will. It is essential that an earth conductor is present in any extension lead. Note, that the PAT Regulations state that there must be only one extension lead between a supply and load and that extension leads must not be plugged into each other. The term used for this practice

Figure 16.4 SWA gland and earth connection "banjo" (by kind permission of Oaklands College, St Albans)

is 'piggy-backing'. If an extension lead is supplied on a wound drum, it should be fully unwound before use so that it does not overheat.

Connections to live parts must be free of mechanical stress. When the flex enters a lampholder, fused switch or plug top, it must be anchored or clamped so that the strain is borne by the flex itself and not its connected cores. When entering an accessory back-box, it should be secured with a compression-type flex gland.

KEY TERM: Fixed wiring – The wiring that forms the installation system.

PVC flex

This is the basic type, which is generally circular and contains two or three cores. Four- or five-core flexes are also available for three-phase systems or where extra control conductors are required. This type of flex is used to:

- Connect domestic and industrial equipment and appliances to the supply
- Hang and connect lamp holders from ceiling roses
- Connect hanging luminaires to the supply

Rubber flex

Constructed from either natural or synthetic rubber, this type of flex is used where a more rugged cable is required. They can withstand:

- Mechanical stress
- Higher and lower temperatures – which makes them suitable for outdoor applications
- Damage from oil and other chemicals

Silicone rubber flex

Manufactured from silicone rubber, this type of flex is suitable for high-temperature conditions. They are used in the food industry because they can both withstand the heat necessary for food production and are also easy to clean of food particles.

Teflon flex

Often used by the chemical and pharmaceutical industries, Teflon flex is resistant to both high temperatures and chemical contact. The insulation and sheath are made from polytetrafluoroethylene (PTFE).

Arctic flex

A PVC flex designed to withstand temperatures as low as $-4\,°C$. It is used for outdoor applications such as power tools and temporary supplies.

SY flex

Although not recognised by the IET On-Site Guide as a type of flex, SY is in widespread use as an armoured flex able to withstand considerable mechanical

Figure 16.5   SY flex (by kind permission of Oaklands College, St Albans)

stress. It usually has a clear outer sheath through which its braided armouring can be seen. Unlike SWA cables, SY armouring is constructed to allow the cable to be flexible. SY requires a specialised gland for termination which enables the armouring to be utilised as an earth conductor (Figure 16.5).

Fireproof cables

During a fire, it is vital that alarms and emergency lighting continue to work for as long as possible. Fireproof cables have been developed to ensure that this happens. Their outer sheaths and conductor insulation are designed to endure high temperatures without burning away. There are two main types of fireproof cable.

Fire performance

Fire Performance (FP) cable is light, flexible and easy to install. It consists of a low-smoke zero halogen outer sheath that will scorch and blister but not burn or emit toxic fumes. Beneath this outer layer is an aluminium sheath. It can serve as a screening as well as a further layer of fire retardant. It also gives the cable the partial rigidity it needs to be straightened, formed into bends, and installed. Typical cable insulation is polyethene (XLPE). FP cables are designed to operate in temperatures up to 950 °C. Some versions of FP are fitted with a silicon-based tape insulation.

The cable can be clipped directly to the surface, using p-clips or run on cable tray. A specialised fire-resistant plastic gland is required to terminate the cable.

Mineral insulated

Invented by Arnold Francois Borel in 1896, mineral-insulated cable was the most widely used fire-resistant cable until the 1990s, when it was superseded by FP, which is much cheaper and easier to install.

Mineral insulated cable consists of solid, bare cores of high-conductivity copper embedded in highly compressed magnesium oxide. The conductors and insulating powder are contained in a solid-drawn copper sheath. This sheath may be further protected by an overall covering of PVC, which will prevent deterioration in the presence of moisture and of moist chemicals (Figure 16.6).

Standard cables are available with one, two, three, four or seven cores and voltage ratings of 500 V and 750 V. Mineral-insulated cables:

- Are non-ageing
- Have high mechanical strength
- Are small in size
- Are very resistant to corrosion
- Have excellent electrical properties
- Are waterproof, can work at high temperatures
- Are suitable for use in flameproof installations

They must be terminated with special seals to prevent the insulator from becoming moist, when it will lose its insulating properties.

The hard-drawn sheath may crack owing to work-hardening if bent too often or too tightly. Usually the minimum bending radius must not be less than six times the cable diameter, although if the cable is bent only once, the minimum radius may be three times the cable diameter.

There is often no identification of the cores of multicore MI cable, which must be checked through with a continuity tester and marked with coloured sleeves. The copper sheath of this cable must be kept separate from the metalwork of water and gas systems or, where this is not possible, must be securely bonded to them. Mineral-insulated cables with aluminium conductors and sheath are also available.

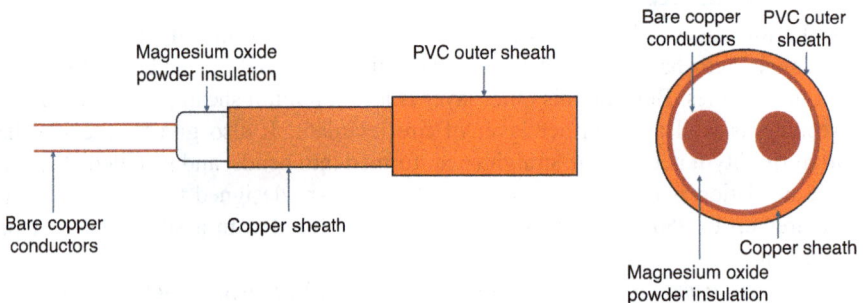

Figure 16.6   Mineral insulated cable

Mineral insulated cable glands

The mineral insulation of these cables is hygroscopic. This means that the insulation will absorb moisture from the air. Wet insulation of this type loses its insulating properties, so the cable ends must be sealed to prevent moisture getting in.

The sealing pots used are screwed or wedged onto the cable sheath and enclose the moisture-resisting compound, which is compressed by crimping a disc into position in the pot end. The disc also secures the PVC sleeves used to cover the bare copper conductors.

Tri-rated cables

Tri-rated cables are designed for use in switchgear and electrical panels. They can operate in temperatures of up to 105 °C, and there are versions with voltage ranges from 0.6 V to 1 kV. Because they are flexible and stranded, tri-rated cables are ideal for panels and machinery where there is limited room and sharp bends are required. The term tri-rated comes from the fact that these cables conform to the standards of three separate countries.

Overhead cables

When running supplies overhead across a space, it is important to use the proper cables. This is because the cables will be exposed to both low and high temperatures and also to UV light from the sun. Also, the cable must be supported along its length and not hang due to its own weight.

Overhead cables should either be supported on or have an integral catenary wire. The integral catenary wire is usually made of steel and while it will conduct electricity, its main purpose is to take the weight of an overhead cable run.

When terminating these cables, the cable is formed around an insulator. A binding wire holds it tight, which means that the actual connection is under no strain.

Bare conductors

Bare conductors are not covered with insulation. Those intended for use at voltages that are above a safe level will normally be supported on insulators. BS 7671 approves the application of bare conductors where they are used as the following.

- High-voltage overhead supply systems where cables are suspended between pylons but out of reach of, what BS 7671 describes as ordinary persons (those with no training or experience in electrical installation).
- The external conductors of earthed concentric systems a type of wiring uses an inner, insulated core for its phase conductor, and the outer conductor or sheath as a combined neutral and earth conductor.
- The conductors of systems working at a safe voltage; for example, where galvanised-steel wire is buried in the soil and fed from a transformer at low voltage to provide soil heating.
- Protected rising-main and busbar systems, which use bare busbars enclosed within a steel trunking and supported on porcelain or plastic insulators.

Provision is made for supplies to be tapped off at intervals. The system is often used as a mains-supply distributor for multi-storey flats or for factories.

- As current-collector wires for overhead cranes and the like. The conductors must then be well out of reach, and clearly labelled to indicate the danger.

Data cables

Coaxial

Coaxial cables are used for audio and visual signals, for example, to connect a television to an aerial and transmit the signals into the set. The cable consists of a copper core insulated by a tough plastic-based layer. A copper braid shields the signal from external interference. An outer plastic sheath provides mechanical protection and holds the components together. When terminating, it is vital that the shielding does not come into contact with the central core (Figure 16.7).

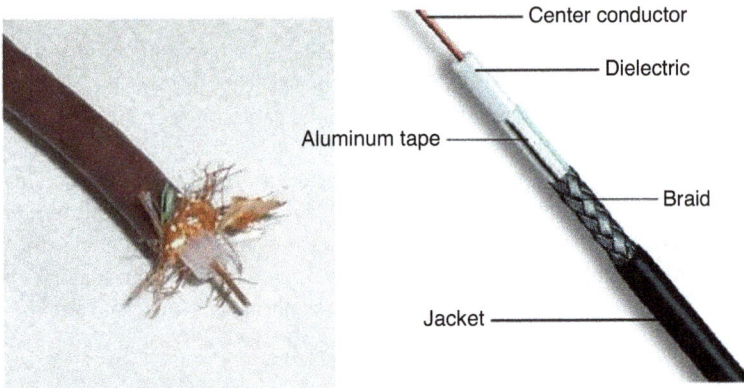

Figure 16.7   Coaxial cable. Image credit: RONALD, via Wikimedia Commons, CC BY-SA 3.0.

Twisted-pair data cables

These are the cables are the backbone of the local area network (LAN) that carries data for computers, routers and telephone systems in commercial premises. The cores are formed into twisted pairs in order to reduce electromagnetic interference and cross-talk between the pairs (Figure 16.8).

Twisted-pair cables are used to connect landline telephones to the telephone company's main connection. These systems are becoming increasingly obsolete with the near-universal use of mobile digital devices, while companies have adopted LAN-based telephone systems.

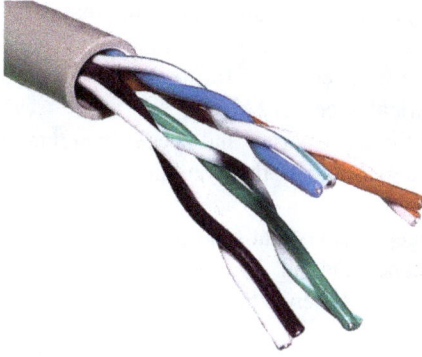

Figure 16.8   Twin-twisted pair cable. Image credit: Baran Ivo, via Wikimedia Commons, Public domain.

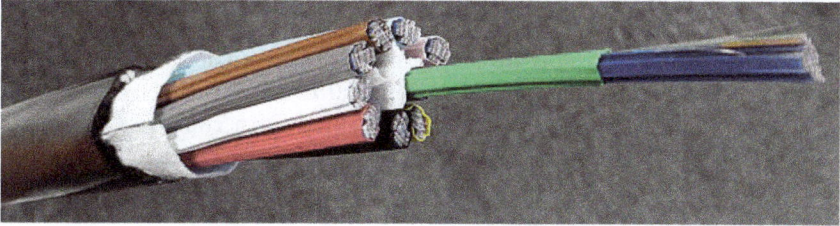

Figure 16.9   Fibre-optic cable

Fibre optic

Fibre optic cables use light pulses to transmit data packages. The cores are manufactured from glass or high-grade transparent plastic. The packages are converted to laser-generated light pulses that travel through the glass cores. Apart from speed, another advantage of fibre optic cables is the fact that several data packages, information, audio, video and voice can be transmitted through a single core simultaneously (Figure 16.9).

Care must be taken when installing fibre optic cable:

- Bends must not be too tight but sweeping, so as not to damage the delicate cores.
- Sharp bends in the cable cause light leakage and weaken the signal
- The cores are fine and sharp, and carry the risk of puncture wounds
- Never look in the cores of a connected fibre optic cable. The laser pulses are bright enough to cause eye damage.

16.4   Wiring systems

Conduit

Conduit is used where a measure of extra mechanical protection is needed for cables. It takes the form of a pipe, through which the cables are drawn. Conduit is fixed on a wall or ceiling using saddles:

- Spacer bar – a simple saddle consisting of a bar that is fixed to the surface and then a moulded top that clamps the conduit in place, using a pair of grub screws. The conduit will need to be formed into a double set to align it with the knockouts in electrical accessory boxes.
- Deep spacer bar – the bar is larger and deeper than that of the spacer bar. This type is manufactured to align the conduit with the knockouts in electrical accessory boxes.
- Hospital – extra deep saddles that hold the conduit off the wall far enough for the wall to be cleaned behind. This would be important in an environment where cleanliness is vital, hence the name *hospital* saddle.

A variety of boxes, couplers and other accessories are available. Conduit can be formed into bends. This should be carried out using a combination conduit bender and vice.

Steel conduit

Both light and heavy gauge steel conduit is available. It can be cut to length and threaded.

Some conduit connections are shown in Figure 16.10. Joints are accomplished by means of couplers (Figure 16.10(a)). Circular or rectangular boxes provide a means for drawing-in cables or mounting accessories. Connections to boxes may be by means of threaded spouts, or using a male bush and coupler to connect to a clearance hole (Figure 16.10(b)).

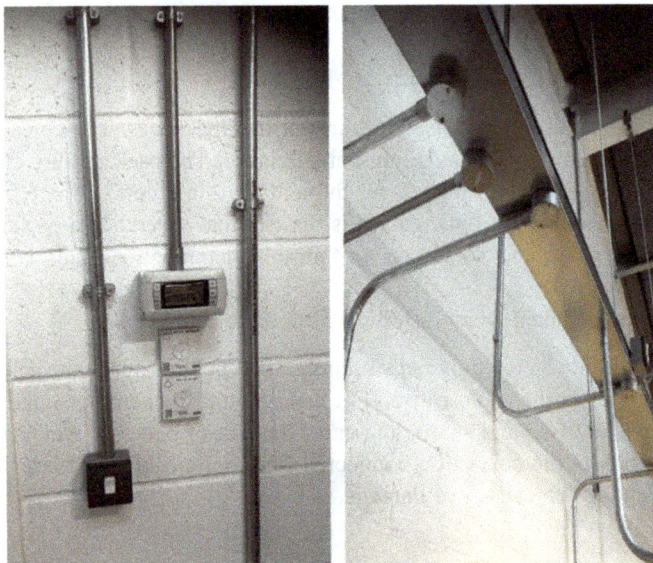

Figure 16.10   (a) Conduit drops to accessories showing couplers and saddles and (b) conduit terminated into circular boxes on steel trunking (by kind permission of Oaklands College, St Albans)

In most industrial applications, there is no need to hide conduit, and it is fixed to the surface of walls and ceilings by means of saddles. To reduce the resistance of the conduit system when used as a protective conductor, to allow for easy drawing-in of cables and to prevent damage to the insulation, conduit ends must be butted up in couplers, and all burrs must be removed.

Conduits are often chosen to accommodate cables for installations in concrete buildings, such as flats and office blocks, and here the installation must be hidden. This is frequently accomplished by burying the conduit in the concrete itself. Special outlet boxes are available for this purpose.

Heavy-gauge conduit is bent where necessary using a bending machine. The bending radius must never be less than 2.5 times the outside diameter of the conduit. Fire barriers must be placed around a conduit where it passes through a wall or floor to prevent the spread of fire.

Plastic conduit

Plastic conduits, usually of hard PVC, are widely available as an alternative to steel conduits. They cannot, of course, be used as circuit protective conductors, but their lightness, cheapness and speedy fixing more than repay the need for an additional protective conductor. They do not have the same mechanical strength as steel conduits and become soft at about $60\,°C$ and brittle below $-15\,°C$ but are increasingly used for both buried and surface work (Figure 16.11).

Joining is by a push fit joint, often used in conjunction with an adhesive. Long runs installed in areas with high ambient temperatures should be fitted with

Figure 16.11   Image of PVC trunking (by kind permission of Oaklands College, St Albans)

expansion couplers. These couplers are longer than standard ones. The pipe is not glued but allowed to slide back and forth inside the coupler as it expands and retracts.

Flexible conduit

Available in a bare metal or plastic-covered finish, or as pure PVC, flexible conduit is used to connect equipment to the supply in a similar manner to flexible cable. It is often used when a large number of conductors are required for the connection. Flexible trunking does not have to be supported as rigid conduit does, but it does require termination glands to secure it to the supply wiring system and to the equipment it feeds.

Trunking

A trunking is a rectangular-section enclosure, available in both steel and plastic. Where large numbers of cables follow the same route, a trunking system will be used instead of a very large conduit, or a multiple run of smaller conduits, because it is lighter, cheaper, quicker and easier to install. Cables can be laid into the trunking before the lids are closed, instead of being drawn in as with conduits and ducts. Special fittings are available for junctions and bends.

Lighting trunking

A specialised trunking is available for large lighting installations. It is either fixed directly to a ceiling or suspended from girders. The trunking hangs upside-down; that is, the open side faces downwards. Clamps or bushes enable fluorescent luminaires to be fixed directly to the trunking and allow wiring to be passed into the light without the need for a flex.

Multi-compartment trunking

In some cases, it is necessary to keep circuits separated. In this case, multicompartment trunking can then be used (Figure 16.12). Circuits are divided into Bands:

- Band I – data and extra low voltage circuits
- Band II – low voltage circuits (230/400 V)

The two Bands are separated by a barrier that creates the trunking compartments.

Outlets for power sockets, telephones, data connections and so on may be mounted on the front of the trunking.

Bus bar trunking

Bus bars were described earlier in this chapter. As well as short bus bar chambers that distribute power among the components of industrial and commercial mains switchgear, it can be pre-fitted into trunking as a power distribution system. An example of its use in a factory where machines are located at a distance from the walls. It would be impractical to run conduits across the floor, especially as many workshops and factories regularly replace or reorganise their machinery. A bus bar system can be run overhead. Machines can be fed via industrial grade flexes connected to the bus bar conductors via take-off boxes.

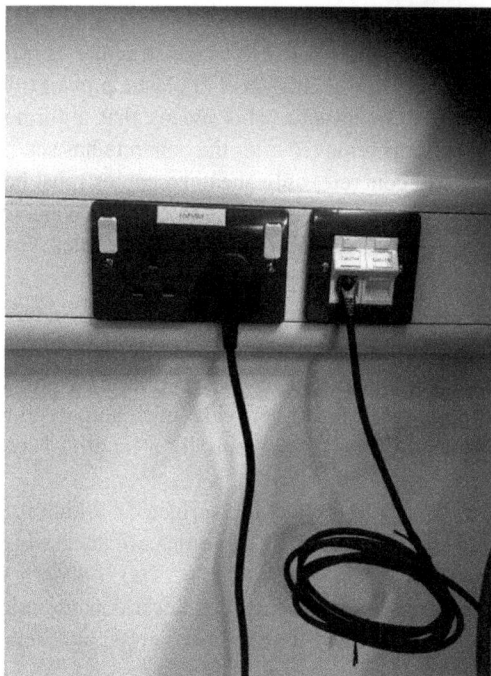

Figure 16.12   *Multi-compartment trunking (by kind permission of Oaklands College, St Albans)*

Some modern lighting can also be supplied using a lighter-gauge bus bar trunking. The lights can be clipped onto the trunking and connect directly to the bus bars.

Mini-trunking

A type of smaller plastic trunking that provides a cable system for surface wiring in situations where **aesthetic** considerations are important. It can be used in a house or office, for example. Mini-trunking is not assembled using fixed joints, rather the ends of trunking lengths are butted together. The angles and tee-pieces clip over the top of the trunking. Unsheathed cables must not be run in mini-trunking. Many types are supplied with double-sided adhesive tape for fixing.

> **KEY TERM: Aesthetic/Aesthetics – Put simply, this is how things look. In other words, is this pleasing to the eye? In electrical installation terms, it is the concern that the components of an installation, conduit, trunking and accessories such as switches and socket outlets are in keeping with the environment they are in and that they are installed neatly. It would *not* be aesthetically pleasing to install metal conduit and trunking in the lounge of a house, for example.**

Ducts

A duct is a closed passageway formed within the structure of a building into which cables may be drawn. It may consist of a steel or plastic pipe or trunking either buried in the floor or ceiling or cast in situ. This means that a former is placed before concrete is poured; when it is removed after the concrete has set, it leaves the duct. If all, or part, of a duct is of concrete, only sheathed cables must be drawn into it.

Ducts are also provided for buried cables to provide ways for easy installation and replacement of armoured, data and other comms cables.

Cable tray

Ideal for steel wire armoured cables and mineral insulated cables, cable tray provides what is termed a wayleave. That is a supported route for cables. Unlike conduit and trunking, tray does not provide mechanical protection. Admiralty pattern is the basic form of tray. It is essentially a length of metal plate, punched with holes for fixings.

It is important to space the tray off the surface on which it is installed. This is so that fixings such as nuts and bolts or cable ties are accessible for tightening.

Right angles and tee pieces are available. Tray benders are also available. Other than good workmanship, aesthetics are not a consideration when installing cable tray. Therefore, it is generally used in commercial and industrial installations.

Ladder

Ladder racking is a heavy duty wayleave. It is constructed in the form of a ladder, with side pieces and cross-pieces. It is designed to carry the weight of large armoured cables.

Basket

The basket resembles a lidless trunking. The sides are not solid but consist of a lightweight struts and frame. It is also available in an upturned V shape. It is generally used for data cabling that is not fixed but laid in the basket. This makes it easier to remove and replace (Figure 16.13).

Figures 16.13   Cable tray and basket (by kind permission of Oaklands College, St Albans)

16.5 Cable clips and cleats

As well as being run through containment or on wayleaves, cables can be fixed directly to a wall or ceiling. BS7671 calls this 'clipped direct'. This doesn't mean that cables can simply be held in place with no concern for aesthetics or the health of the cable itself.

- The work should be neat.
- The cable should be straight.
- Clips and cleats should be the correct, and a uniform, distance apart.
- Cables must not be run diagonally but in straight lines.
- Cable bends should not exceed the minimum radius to avoid damage to the cores and insulation.
- Clips or cleats must not be present on the curve of a cable bend.

The IET On-Site Guide gives maximum distances between cable fixings for various cable types and for vertical and horizontal runs. It also gives formulae for the radii of cable bends.

Figure 16.14 shows some of the different types of cable clip and cleat available.

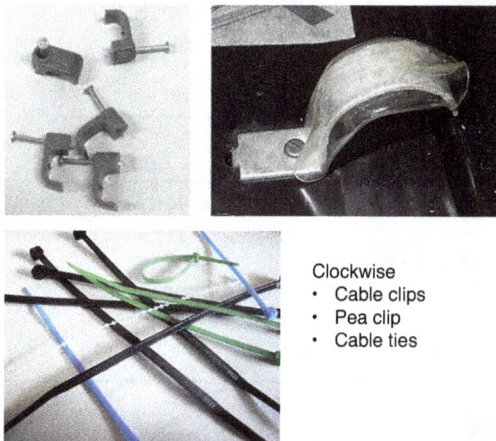

Clockwise
- Cable clips
- Pea clip
- Cable ties

Figure 16.14   Types of cable fixing (by kind permission of Oaklands College, St Albans)

16.6 Conduit and trunking capacities

The number of cables that can be fed into a conduit or laid in trunking will be limited by both physical and electrical considerations. Insulation can be damaged if too many cables are crammed into a conduit.

The other issue is heat. We have already seen that a magnetic field forms around current-carrying conductors. Because magnetism is a form of energy, it will eventually transform into heat. If cables are compressed together with no means for that heat to dissipate, their conductor resistance will be correspondingly higher. This will generate even more heat and could compromise the insulation and potentially cause a fire.

Appendix E of the IET On-Site Guide features a set of tables from which minimum sized conduits and trunking can be calculated. The calculation is carried out as follows.

1. Select correct table.
2. Each cable size (1.5 mm², 2.5 mm² 4.0 mm², etc.) is given a factor.
3. Multiply the number of cables of each size by their factors.
4. Add the total factors.
5. Use this figure to find the minimum size conduit or trunking using the appropriate table.

Example 16.1 (Factors taken from IET On-Site Guide)
The following cables are to be run in a 3 m length of conduit that incorporates two bends?

7 × 2.5 mm²	– Factor 30
3 × 6 mm²	– Factor 145
7 × 30	= 210
3 × 145	= 435
Total factors	= 645

Conduits factors are:

20 mm	233
25 mm	422
32 mm	750

750 is the first factor to exceed 645;
therefore, selected conduit is 32 mm

16.7   Cable terminations

Once the cable is wired to an accessory or to electrical equipment, it has to be terminated securely. It is vital that the connection is tight because any movement can cause an arc that leads to overheating and even fire. There are a number of different types of termination, and these are described below.

Screw-type

For many years this was the standard method for connecting conductors to accessories such as light switches, isolators and 13 A socket outlets. The conductor is pushed into the terminal and a grub screw is tightened to secure it in place. More than one conductor can be accommodated into this type of terminal. Care must be taken to ensure:

- The conductor(s) is/are secure.
- No bare copper is exposed.
- The grub screw is not pressed onto insulation rather than the conductor itself.
- The screw is not overtightened.

Post type

Often used to terminate larger cables and for appliances such as ovens and hobs. The post is a threaded stud. The conductor is secured using nuts and washers. The most effective method of terminating a cable onto a post-type termination is to use a lug. If, however, a bare conductor is to be connected, then it should be wrapped round the stud in a clockwise direction, the same as that for tightening a nut. This prevents the conductor from being unwound as the nut is tightened.

Push-fit

Accessories and connectors are now mostly fitted with push-fit terminals. There have been many attempts to replace screw-type with these over the years but a reliable system has finally been introduced. To secure a conductor, a small lever is raised, the conductor is inserted in the terminal and the lever is pushed down again.

Most terminals will only accept a single conductor so accessories are fitted with sets of linked terminals, e.g. three line, three neutral and three earth terminals.

Punch type

Used mainly for small conductors such as twisted-pair. These wires are extremely thin and snap easily when being stripped for connection. The punch-type terminal is V-shaped with a pair of blades at the bottom. A specialist tool is used to punch the cable core down into the vee, where the blades cut through the insulation and form the connection.

Lugs

There is a wide variety of lug types, from those suitable for normal hard wiring cable sizes (1.5 mm^2, 2.5 mm^2, etc.) to larger versions designed for use on supply cables.

In the past, lugs were secured on the conductor using hot metal, and although this type of lug is still in existence, the majority are compression type. A crimping tool, graduated to the correct setting, should be used to secure a lug without crushing the conductor inside (Figure 16.15).

Soldered

Soldering is, in effect, a form of micro-welding. Heat is applied to both the terminal and conductor, and at the same time, the solder material is melted onto them to secure the conductor in place.

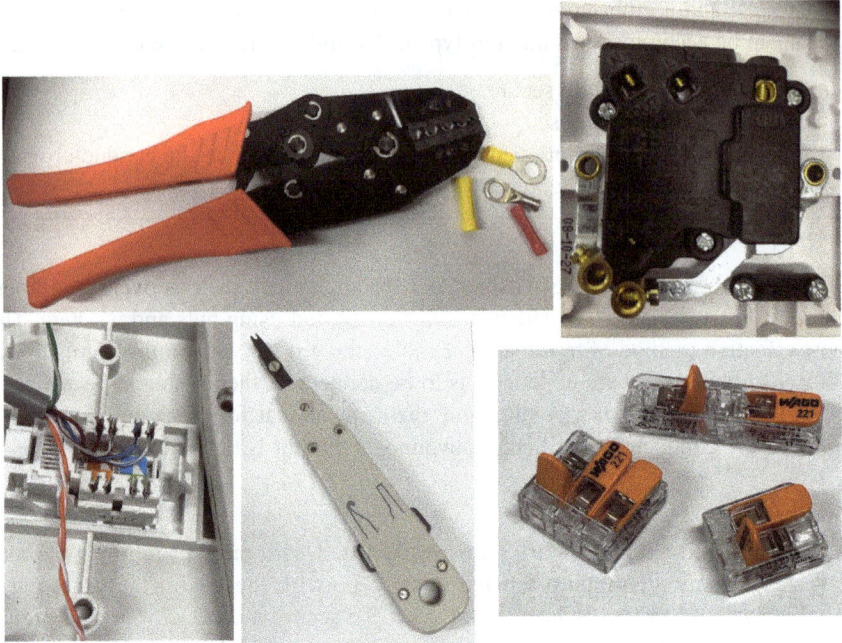

Figure 16.15   Crimping tool and lugs, screw-type terminals etc. (by kind permission of Oaklands College, St Albans)

Particularly suitable for the fine conductors used on electronic circuit boards, care must be taken not to damage nearby components or the insulation of the cable being soldered. Remember, the soldering iron is hot, also the fumes from the solder is harmful. Good ventilation is vital when carrying to this operation.

16.8   Exercises

1. List the materials used as conductors for power cables, and compare their advantages.
2. Give the names of four materials used to insulate cables, and state where each could be used with advantage.
3. The following materials are used in the manufacture of cables. By giving the important properties of each, show where it could be used.
    (a) copper
    (b) aluminium
    (c) PVC
    (d) PCP
    (e) paper
    (f) glass fibre

4. List two situations in which bare conductors may be used.
5. Describe the construction of a twin-with-protective-conductor sheathed wiring cable. What are the advantages of this type of cable, and in what type of installation it likely to be used?
6. Using a sketch, describe the construction of a mineral-insulated cable. Show how this cable is terminated, and explain why a special termination is necessary.
7. List What is FP cable generally used for? Describe its construction.
8. Describe the construction of a PVC-insulated PVC-sheathed steel-wire-armoured cable.
9. Describe three termination types.
10. What is ladder tray?
11. When would flexible conduit be used?
12. What is multi-compartment trunking and why is it used?

16.9   Multiple-choice exercises

16M1 How is a cable conductor manufactured?
  (a) Sweated
  (b) Molten metal in a mould
  (c) Cold drawn
  (d) Drag drawn

16M2 The most usual conductor material for cables is
  (a) aluminium
  (b) sodium
  (c) silver
  (d) copper

16M3 Aluminium is often used as conductor material on heavy power cables rather than copper because
  (a) it is cheaper and lighter
  (b) it is likely to corrode more easily
  (c) it will work harden and break when repeatedly bent
  (d) it can be safely fixed with copper or steel saddles

16M4 A flexible cord or cable will have a conductor made up of more strands than a fixed wiring cable of the same cross-sectional area because
  (a) it is easier to make it that way
  (b) if soldered, the conductor may lose its flexibility
  (c) the strands can be twisted together before being connected
  (d) it will bend more easily

16M5 A disadvantage of the common insulation material PVC is that
  (a) it is extremely expensive
  (b) it has a complicated name

(c) dense smoke and corrosive fumes are emitted when it burns

(d) it can be extremely difficult to strip

16M6 A cable insulation that is resistant to heat, water, oil and sunlight; and may be buried directly in the ground and exposed to weather is

(a) ethylene–propylene rubber

(b) polychloroprene (PCP)

(c) magnesium oxide

(d) cross-linked polyethylene (XLPE)

16M7 The insulating material used in mineral-insulated cables is

(a) PVC

(b) magnesium oxide

(c) sulphur dioxide

(d) hard-drawn copper

16M8 An armoured cable is likely to be chosen for installation

(a) in the kitchen of a normal house

(b) overhead between a house and a shed

(c) of a temporary nature that will be replaced within three months

(d) in a heavy industrial situation where there is a danger of mechanical damage

16M9 An insulator that absorbs moisture from the air is called

(a) hygroscopic

(b) cross-linked polyethylene

(c) absorbent

(d) wet

16M10 One reason for limiting the number of cables drawn into a conduit is

(a) so that the cables are in very close contact with each other

(b) to limit the cost of the installation

(c) to ensure that cables can be easily installed and removed

(d) so that the conduit does not become crowded

16M11 Heavy-gauge steel conduits are joined together by

(a) push-fit connectors and adhesive

(b) welding or brazing

(c) threading and threading couplers

(d) butting firmly together

16M12 Where large numbers of cables are concerned, the use of trunking rather than conduit is an advantage because

(a) trunking is rectangular in section rather than circular

(b) cables can be laid in rather than having to be drawn in

(c) the cables can be identified more easily

(d) bends can be tighter in trunking than in conduit

16M13 Different categories of circuit can be run in a common trunking if
  (a) they are separated from each other by compartments
  (b) they are each insulated for the voltage they carry
  (c) they are taped together
  (d) they are separated within the trunking by at least 10 mm

16M14 What happens to a cable when its temperature increases?
  (a) it will be difficult to find if a fault develops
  (b) the heat will attack the cable's sheath and insulation
  (c) it will become more efficient at carrying current
  (d) its resistance will rise, its current carrying capacity will be reduced

16M15 Where will the maximum distance between cable fixings and minimum bend radii be found?
  (a) IET On-Site Guide
  (b) BS7671
  (c) IET Guidance Note 3
  (d) GS.38

Chapter 17

Electric cells and batteries

17.1 Storing electricity

Although renewable energy systems such as wind and solar (photovoltaic) make a significant contribution to the nation's commercial electrical supply, most electrical power is generated in power stations by rotating machines that stop producing voltage and current the moment they cease operation. A large number of generators are necessary to provide the power required by the grid. This maximum load is not required constantly, however, but at peak times. This means that there are periods when many generators stand because they are needed only to meet peak demands.

Fewer generators would be necessary if electricity could be stored during the night for daytime use. This is achieved, to a certain extent, by a storage system in which water is pumped up to a high-level reservoir during the period of low demand at night and then used to drive hydro-turbines at times when demand is high. Lack of suitable sites prevents the system from being widely applied.

Another method of storage is provided by a chemical process, which can be converted to electrical energy as required. Although the cost of such an operation on a national scale would be prohibitive, this method is very widely used for individual systems and electrical devices. The component used for chemical-to-electrical energy conversion is called a cell.

Cells and batteries

It is important to clear up any confusion regarding the terms cell and battery.

- A cell is a component that produces electricity. It is the beating heart of a battery. A cell has a fixed voltage, depending on its constituent chemicals and materials.
- A battery is a collection of cells. The voltage output of a battery is the sum of the voltages produced by its cells.

The main part of this chapter will be focused on individual cells. There are two types:

- Primary
- Secondary

Primary cells

Primary cells have the active chemicals placed in them when they are made. When the chemicals are used up, they are sometimes replaced, but it is more usual to throw away the spent cell. Most primary cells are now made in the dry form and are widely used in torches, toys and digital watch batteries.

Secondary cells

Secondary cells are capable of being reactivated when their energy is spent. This is achieved by passing a charging current of electricity through them. Some of this energy is converted to the chemical form and stored for future use. Secondary cells and batteries are widely used to power mobile devices, tools, small portable equipment on vehicles and for standby supplies in case of mains failure.

Basic components of a cell

Although cells have increased in sophistication and efficiency over the years, the basic components and principles of operation remain the same as earlier, more primitive, types.

- Positive and negative terminals – the two terminals from which the voltage generated by the cell can either be transferred to the load or the cell recharged.
- The electrolyte, which is usually an acidic solution, allows ions to flow between the anode and cathode. This movement of ions is the source of the electrical charge provided by the cell. Electrolytes can be a liquid or a paste.
- Cathode – the cathode is considered to be the positive terminal when the cell discharges into a load. Negatively charged electrons migrate to the cathode through the electrolyte. However, if the battery is rechargeable, the cathode becomes the negative pole during the recharging process.
- Anode – the negative terminal. Oxidation of the anode, created by the action of the electrolyte, releases electrons that flow to the cathode. However, if the battery is rechargeable, the anode becomes the positive pole during the recharging process.

   KEY TERM: Ions – Electrically charged molecules or atoms, an effect caused by an unequal number of protons and electrons.

17.2   Primary cells

Primary cells are those which have their energy added in chemical form during manufacture and which normally cannot be recharged once this energy is spent. The easiest way to understand primary cells is to first take the simplest of them, examine its defects and then show how these are overcome in more complex cells.

Simple cell

Figure 17.1 shows a simple cell, which consists of plates of copper and zinc immersed in a weak solution of sulphuric acid. This acid solution is the electrolyte.

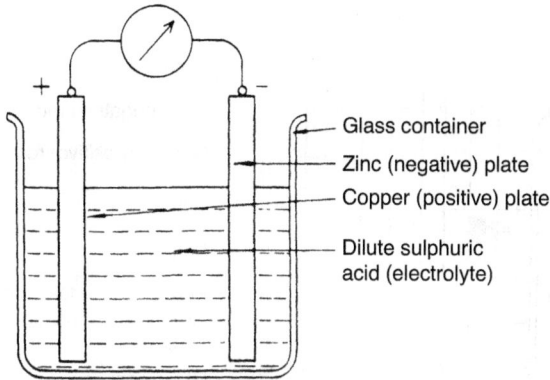

Figure 17.1   Simple electric cell

If an external circuit is connected across the plates, current flows from the copper (cathode – positive) plate to the zinc (anode – negative) plate. The circuit is completed through the electrolyte.

Hydrogen is produced in this action and collects on the copper plate in the form of fine bubbles, which effectively insulate the plate from the electrolyte. This effect is called polarisation, and, when a current is drawn, results in a sharp decrease in cell EMF from its initial value of 1.08 V.

The zinc plate is not pure zinc but is impregnated with small particles of other metals such as iron and lead. When the plate is immersed in the electrolyte, these impurities, in conjunction with the zinc, form tiny cells on its surface, which erode the zinc plate. This effect is called local action.

These two disadvantages make the simple cell unsuitable for practical use. Practical cells use pure, or mercury-coated, zinc plates to prevent local action. Polarisation is overcome by placing the positive electrode in a chemical that absorbs the hydrogen and is called a depolariser.

Wet Leclanché cell

Cells of this type are no longer in operation and have been replaced by dry cells, which work on the same principle but are more practical and portable. Leclanché cells were classically used for electric bells and telegraph systems.

A section of a typical cell is shown in Figure 17.2.

- The mercury-coated zinc (negative) rod is enclosed within the glass jar and immersed in its electrolyte of ammonium chloride, sometimes called sal ammoniac.
- The carbon (positive) rod is packed in a depolariser of crushed carbon and manganese dioxide, separated from the electrolyte by a porous pot.
- The depolariser is efficient, but rather slow in action, which means that the cell is suited for intermittent operation.

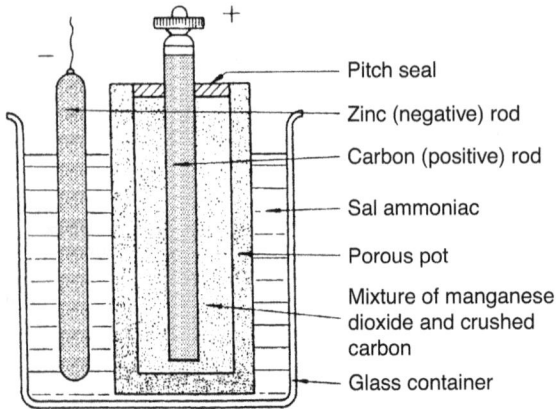

Figure 17.2   Leclanché cell (wet)

- The electrolyte has a tendency to creep up the sides of the glass jar. This is prevented by painting or greasing the glass.
- The EMF of this cell is 1.5 V.

Zinc chloride cell

This is a form of Leclanché cell in which the liquid is replaced by a paste that consists of zinc chloride mixed with water and ammonium chloride. These batteries can be small and portable and used in any position. The sheet-zinc container acts as the negative pole. The positive is provided by a carbon rod held inside the electrolyte. The zinc tube, often slid into a cardboard tube, holds the electrolyte (Figure 17.3).

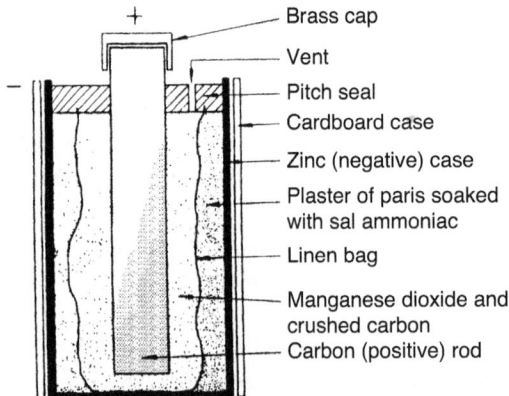

Figure 17.3   Zinc-chloride cell

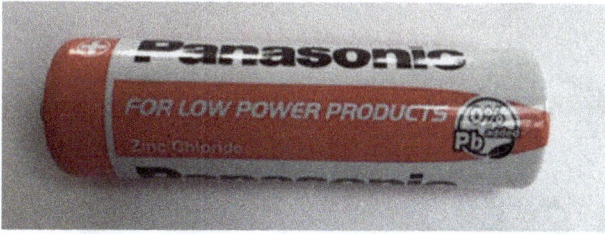

Figure 17.4   Modern zinc oxide battery

The cell EMF is 1.5 V when new, falling to a steady 1.4 V in service. When the cell voltage falls to 1 V, it should be discarded, since at this stage the zinc container quickly corrodes and allows its contents to escape. Leakproof cells enclosed in a second steel case are available to prevent this hazard (Figure 17.4).

Alkaline cells

Alkaline batteries are in common use and ideal for household devices. One advantage of the alkaline cell is its long shelf life, which can be years. This makes them a good choice for equipment that needs standby power. They are also relatively cheap and are available in a number of sizes. The alkaline manganese cell has a voltage of 1.5 V and has a number of useful advantages, including:

- self-venting to prevent bursting in the event of a sustained short circuit
- a much more uniform low internal resistance
- can provide comparatively heavy current for long periods without the need for a rest period to recover
- will operate satisfactorily over a very wide range of ambient temperatures (typically -20 °C to $+70$ °C).
- very good leakage protection

The construction of a typical alkaline cell is shown in Figure 17.5.

Lithium cells

With a range of voltages from 1.5 V to 3.7 V, a single lithium cell is often able to replace two or more cells of other types. They have exceptionally long shelf and operating lives, sometimes in excess of ten years, so they are used in equipment where the frequent need to change batteries would be a problem, such as cameras and portable computing devices. They are available in a variety of sizes or forms, including the small 'button' type used for devices such as digital watches and car keys.

Their ability to store energy is almost three times that of other cells of the same size, they have a much wider operating temperature span and they have a much greater ability to provide high currents. The anode is made of lithium and cathode from a material, such as lithium manganese oxide or cobalt oxide, that will readily

Cathode cap

Insulating washer

Outer steel jacket

Separator

Anode of powdered zinc

Electrolyte of potassium hydroxide

Cathode of mixed manga dioxide and graphite

Cathode collector

Anode collector

Plastic grommet

Vent

Insulator

Anode cap

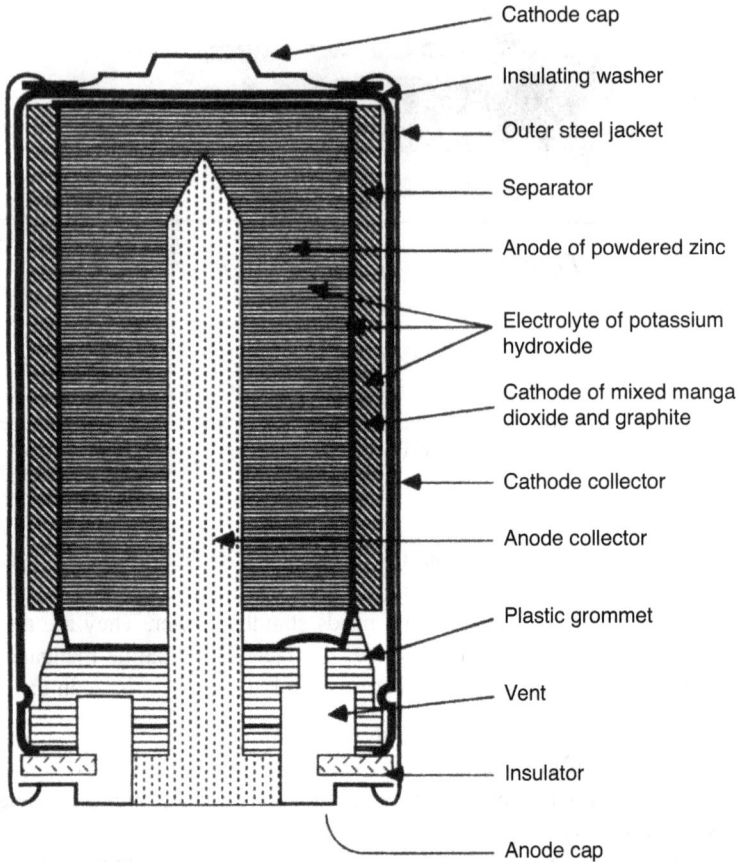

Figure 17.5   Diagram of an alkaline cell

accept electrons. In some of the most advanced lithium cells, thionyl chloride is used for the anode, cathode and electrolyte. Further advances are being made into solid, rather than liquid-based electrolytes, with ceramics being one of the materials under development. They are also available as re-chargeable batteries.

Silver oxide

Available mostly as button batteries due to the high cost of silver, this type of cell uses silver oxide as the cathode and zinc as the anode. Advantages of this type are a high-energy output and a long life. Their output voltage is 1.6 V. A disadvantage is the inclusion of mercury in the cathode, a metal considered to be a highly toxic pollutant. Whilst the button versions (Figure 17.6) are cheap, larger silver oxide batteries are used by organisations such as the armed forces and in the space programme.

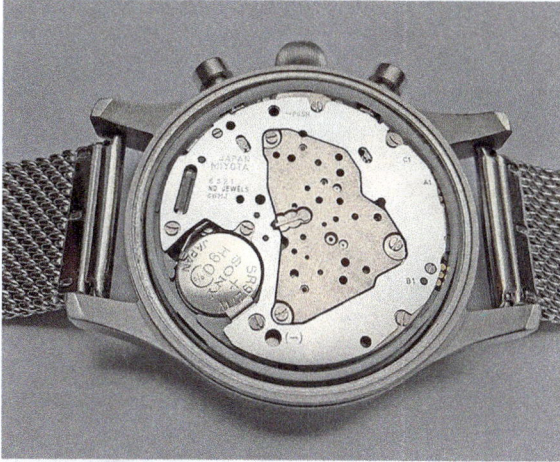

Figure 17.6   Typical use for a silver oxide cell(s). Image credit: Francis Flinch, via Wikimedia Commons, CC0, Public Domain.

17.3   Secondary cells

There are four basic types of secondary cell in general use:

- Lead–acid
- Alkaline
- Nickel Cadmium (Ni-Cd)
- Lithium Ion (Li-ion)

A secondary cell has a reversible chemical action, which means that it can be recharged by passing an electric current into the cell via its positive and negative terminals. The characteristics, maintenance and charging of secondary cells are covered in Section 17.4.

Lead–acid cells

Lead–acid cells are the most widely used type of secondary cell and consist of two plates of lead in an electrolyte of dilute sulphuric acid (flooded type) (Figure 17.7).

- The positive plate is often in the form of a lead–antimony alloy lattice into which active lead oxide is pressed.
- The negative plate is usually of pure lead.

Water is produced on discharge and lowers the specific gravity of the dilute sulphuric acid. Specific gravity is a method of indicating the strength of the electrolyte, and is the ratio of the mass of a given volume of the electrolyte to the mass of the same volume of water.

Measurement of **specific gravity** using a hydrometer (see Figure 17.8) is a good method of determining the state of charge. If the cell is over-discharged or

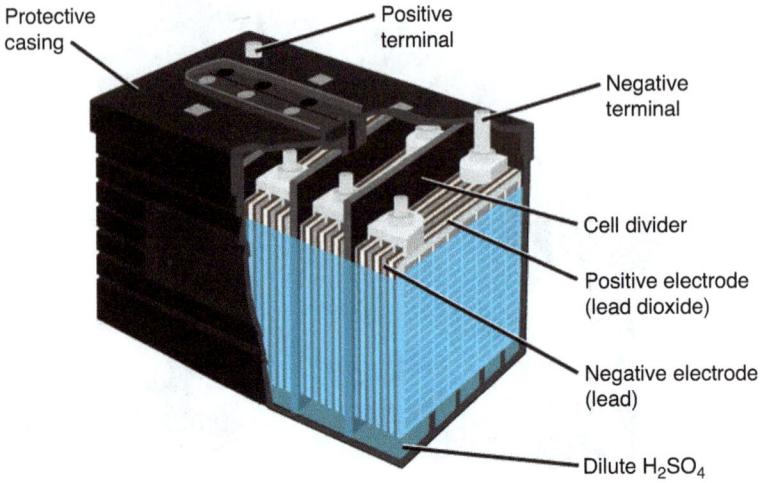

Figure 17.7   Modern lead acid cell. Image credit: OpenStax, via Wikimedia Commons, CC BY 4.0.

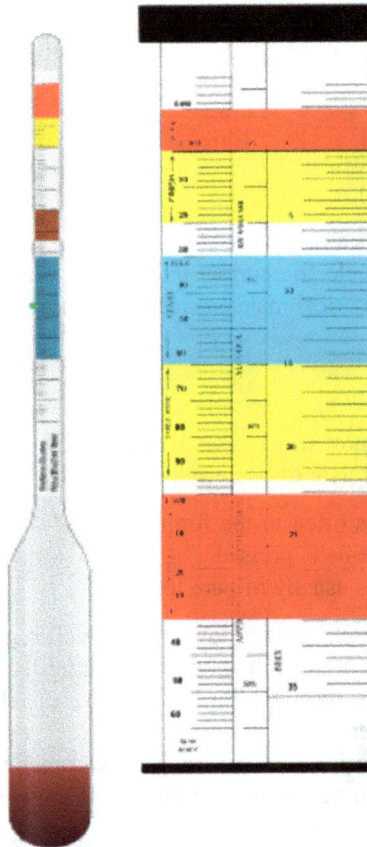

Figure 17.8   Modern hydrometer. Image credit: SG0039, via Wikimedia Commons, CC0, Public Domain.

allowed to stand for long periods in the discharged condition, the lead sulphate coating on the plates becomes hard and is difficult to remove by charging. In this condition, the efficiency of the cell is reduced, and it is said to be sulphated. A healthy lead–acid cell has an initial EMF in the region of 2.2 V, which falls to 2 V when in use.

KEY TERM: Specific Gravity – The density of a substance in relation to the density of water. Readings can be taken using a hydrometer.

The construction of lead–acid cells varies according to intended use. A common application is as car batteries. These consist of a set of cells, each in a separate container held together in a box made of a hard-wearing plastic. Terminals for each cell are brought through the top of the moulding. Unsealed types have removable vent plugs for inspection and topping up. To obtain maximum capacity for minimum volume, the plates are mounted close together but prevented from touching by porous plastic separators.

Stationary cells are permanently installed in buildings to provide power for systems such as emergency lighting and alarms, which must operate when the mains supply fails. Uninterrupted Power Systems (UPS) can be fitted with lead-acid cells, which are on a permanent charge until a power failure, when they provide short-term emergency cover until power is restored or a standby generator comes online.

Since stationary cells are unlikely to be subjected to mechanical shocks, the containers are usually in the form of glass or plastic containers, with active plates and separators suspended within them.

All lead–acid cells have space left below the plates for accumulation of active material forced out of the plates while in use, which might otherwise 'short' out the cell (Figure 17.7). The specific gravity of a charged cell varies with its type and is also slightly affected by temperature, but an average value is 1.25. All lead–acid batteries pose dangers to the maintainer, and these are considered in Section 17.4.

Conventional lead–acid cells emit oxygen and hydrogen during operation, the gases being replaced by 'topping up' with water (H_2O). Modern types are 'gel' batteries, containing recombination plugs to convert the hydrogen and oxygen in the cell to water, so that no topping up is required. This allows the electrolyte to be bonded in a **thixotropic gel** rather than a liquid. It is necessary for the cell to be provided with a valve to allow escape of excess hydrogen that results from short-circuit conditions and to prevent the entry of oxygen, which would result in internal corrosion.

KEY TERM: Thixotropic gel – A gel that will thin when disturbed and then re-thicken if left to do so. A common use is in tomato ketchup. Shaking and hitting the bottle will thin the sauce so that it can be poured.

Alkaline cells

There is little difference outwardly between non-rechargeable and rechargeable alkaline cells. The positive, or cathode, terminal is made of steel while the negative anode is formed of powdered zinc suspended in a gel solution. Additives, such as barium, improve cycling and increase cell capacity. Other additives in both the anode and cathode materials reduce the production of hydrogen and prolong the cell's life.

Nickel–cadmium cells (Ni-Cd)

These cells are enclosed in plastic or steel cases and use potassium hydroxide as an electrolyte. The cell has interleaved flat plates:

- Positive – nickel hydroxide
- Negative – cadmium

The robustness of nickel–cadmium cells and their resistance to both mechanical and electrical ill treatment make them suitable in the form of batteries for:

- some types of electric vehicle
- military use
- operation of high-voltage switchgear
- railway signalling
- all situations where failure can have safety implications

These cells will maintain their charge for very long periods without attention and can operate over a range of temperatures which would make other types ineffective.

Small sealed nickel–cadmium rechargeable cells are being used increasingly in both their normal and button configurations because they are small, light and extremely reliable. Some types have the ability to be recharged (Figure 17.9) very rapidly (within minutes). They:

- Have a very long life
- Are maintenance free
- May be overcharged without ill effect

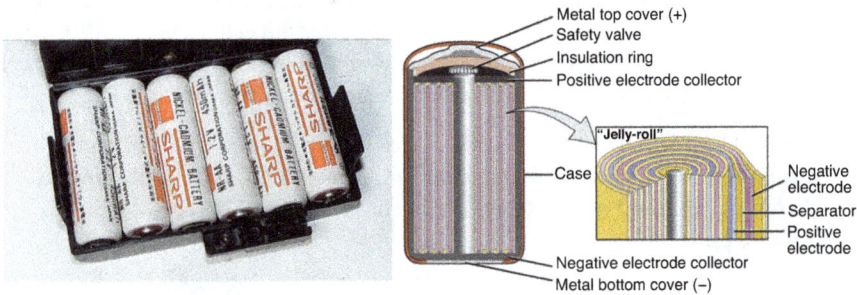

Figure 17.9   Modern nickel cadmium battery. Image credit (right side image): OpenStax, via Wikimedia Commons, CC BY 4.0.

- Can be stored for long periods in either charged or discharged condition,
- Have a high discharge rate and constant discharge voltage
- Are mechanically very robust

These advantages make them the usual choice for applications such as

- Cordless tools
- Camcorders
- Mobile telephones,
- Laptops

Nickel–cadmium cells are usually float-charged during use at a constant voltage, set to minimise overcharge current. This is done to keep them in optimum condition and to avoid the need to have a larger battery to take into account self-discharge.

Lithium ion (Li-ion)

Developed in the 1960s and 1970s, lithium-ion cells experienced a number of setbacks due to safety issues. The electrolyte was flammable, which made them a fire risk. However, more recent advances in battery technology have enabled a safer version to be manufactured.

When under charge, lithium ions migrate from the lithium-cobalt oxide cathode to the graphite anode through an electrolyte made form a lithium-salt solution.

Lithium-ion batteries are used for mobile devices, laptop computers and electric vehicles (Figure 17.10). They perform well at higher temperatures, although prolonged exposure to heat will reduce cell life. They also give good charging performance at cooler temperatures.

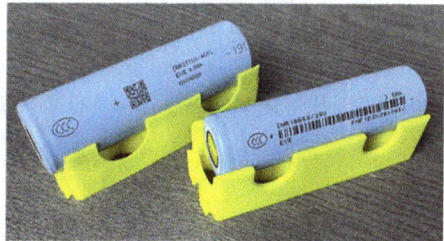

(a)   (b)

Figure 17.10   Modern lithium-ion battery. Image credits: (right side image) Raimond Spekking, via Wikidata, CC BY-SA 4.0 and (left side image)Sevenethics, via Wikimedia Commons, CC0, Public Domain.

17.4   Care of secondary cells

Charging

When the chemicals of a cell have changed to the inactive form, they can be made active once more by passing a charging current through the battery in the opposite direction to the discharge current. The supply voltage must be in excess of the battery voltage, or no charging current can flow. There are two methods of charging:

Constant-voltage charging

A constant voltage is applied to the battery under charge. While the battery is charging, a steady increase occurs in its terminal voltage, so that the charging current tapers off to a lower value at the end of the charge than at the beginning.

Constant-current charging

This system uses either an adjustable voltage source or a variable resistance so that the charging current can be kept constant throughout the charge. Healthier stationary batteries result from this method, but the variation concerned normally requires manual adjustment, although automatic means can be provided.

It should be appreciated that cells and batteries are very temperature conscious and that correct charging levels will change if the battery temperature is unstable.

Modern electronic battery chargers are capable of monitoring the condition of a battery from its voltage on charge and will ensure that the device is not over- or under-charged. In many cases, it may be left connected permanently, recharging the battery after use and maintaining it in a fully charged state at other times. In many cases electronic systems have replaced transformers, so that chargers have become much smaller and lighter.

Stages of charging

In the construction and engineering industry, most handheld power tools are extra-low-voltage battery type, using rechargeable power packs. The chargers are provided with each tool and are lightweight and portable. These chargers are three-stage or multistage chargers. The three stages are called:

- Bulk – 80% of capacity. The current remains at the same level while the voltage increases
- Absorption – voltage stabilises while the current is steadily reduced. This is the longest stage as a slower charge is needed to prevent damage to the battery
- Float – a small, or 'trickle' top-up charge that keeps the battery topped up until it is used.

Rechargeable nickel- and lithium-based battery chargers use the three-stage system. Some cheaper chargers monitor charge levels through battery temperature. This is not a reliable method.

Capacity

The total charge which a battery or cell will hold is measured in terms of the current supplied × the time for which it flows.

In practice, this figure varies depending on the rate of discharge. A quick discharge gives a lower figure than a slow one. Capacity is usually measured at the ten-hour rate; for instance, a 60 ampere-hour (Ah) battery will provide six amperes for ten hours.

Charging and maintenance of lead–acid and alkaline cells are quite different and will be considered separately.

Lead–acid cells: maintenance

These cells are the most widely used of the secondary cells, owing to their comparatively low cost and higher voltage per cell. They can be damaged, however, by charging or discharging too quickly, overcharging, leaving them in the discharged state, etc. Healthy cells can only be maintained in condition either by keeping them fully charged or by periodically recharging, ideally at monthly intervals. A lead–acid cell will lose its charge if left standing over a period of a few months.

Determining charge

For periodic charging or for recharging after use, the constant-current method is generally applied. The current value needed varies with the type of cell. A common value is:

One-tenth of the ampere-hour capacity at the ten-hour rate; that is, 6 A for a 60 Ah battery.

It is important not to overcharge. There are three methods by which the state of charge can be determined.

- Colour of plates – Fully charged cells should have clear light-grey negative plates and rich chocolate-brown positive plates.
- Terminal voltage – Open-circuit voltage of a fully charged cell depends on the type and is within a range of 2.1 V to 2.3 V. If this voltage is measured with the charging current flowing, it will be increased by the voltage drop in the internal resistance of the cell.
- Specific gravity – This is the best method of determining the state of charge. The specific gravity of the electrolyte varies from 1.25 for a fully charged cell to about 1.17 for a discharged cell. These figures apply to flooded-type storage batteries and may vary for other types of cell.

Trickle and float charging

A lead–acid cell can be kept in a healthy condition for long periods by making good the losses due to self-discharge as they occur. This is done by a low current, continual, or trickle charge.

Some DC systems use 'floating' batteries. These are connected directly across the supply and the same nominal voltage. If the battery voltage falls, the greater supply voltage charges the battery until values are equal. If the supply fails, the battery takes over. As mentioned before, this system is the basis of the UPS as well as for non-maintained emergency lighting.

Precautions

Precautions must be carefully taken where lead–acid cells are used or charged, mainly because the cells emit hydrogen and oxygen during charge and discharge. They also contain acid which can damage clothing and burn skin.

Alkaline cells: maintenance

These cells are lighter than the lead–acid type. They also:

- Have greater mechanical strength
- Can withstand heavy currents without damage
- Give off no corrosive fumes
- Are not affected by being left in the discharged condition.

Charge is held much better than with lead–acid cells; in fact, manufacturers claim 70% capacity after three years without attention. Trickle charging is thus seldom necessary, although it can be employed when required. Charging after use, or at six-monthly intervals, is all the attention usually required.

These cells are not damaged by overcharging. Constant-voltage or constant-current charging can be used. Topping up with distilled water is necessary. Internal resistance of these cells is generally higher than the equivalent values for lead–acid cells, so voltage varies over a wider range with load.

17.5   Internal resistance

The path taken by current as it passes through a cell (Figure 17.11) will have resistance. This is the internal resistance of the cell and is important because of the voltage drop which it causes in the cell. This results in the terminal voltage being less than the EMF on discharge. The internal voltage drop, given by multiplying internal resistance by current, must be subtracted from the EMF to find the output voltage, that is,

$$E - IR_c = U$$

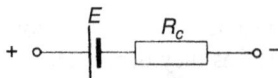

Figure 17.11   Representation of cell in circuit diagram

where E = cell EMF, V; I = current taken from the cell, A; R_c = cell internal resistance, Ω; and U = cell terminal PD, V.

The internal resistance of a cell usually increases with its age, and with ill treatment such as excessive discharging current, standing in the discharged condition and so on.

Example 17.1
A cell has an internal resistance of 0.02 Ω and an EMF of 2.2 V. What is its terminal PD if it delivers

(a) 1 A
(b) 10 A
(c) 50 A?

(a) *1 A*

$$U = E - IR_c = 2.2 - (1 \times 0.02) = 2.2 - 0.02 = 2.18 \text{ V}$$

(b) *10 A*

$$U = E - IR_c = 2.2 - (10 \times 0.02) = 2.2 - 0.2 = 2 \text{ V}$$

(c) *50 A*

$$U = E - IR_c = 2.2 - (50 \times 0.02) = 2.2 - 1.0 = 1.2 \text{ V}$$

This example illustrates how the terminal PD falls off as the current increases.

The internal resistance of a cell depends on its design, construction, age and condition. Internal resistance can be measured using a high-resistance voltmeter and an ammeter connected with a switch and resistor. When the switch is open, no current is taken from the cell (if we neglect the voltmeter current), so the voltmeter reads cell EMF. If the switch is closed, the terminal voltage and current are measured. Since

$$U = E - IR_c; \quad \text{therefore} \quad R_c = \frac{E - U}{I}$$

The resistor R is used to adjust the cell current to a convenient value that will simplify the calculation.

Example 17.2
The EMF of a cell is measured as 2.1 V, and its terminal PD as 1.9 V when it carries a current of 5 A. What is its internal resistance?

$$R_c = \frac{E - U}{I} = \frac{2.1 - 1.9}{5} = 0.04 \ \Omega$$

The symbol for a cell is shown in Figure 17.12. The positive connection is represented by a long line and the negative by a shorter, thicker line. It is often convenient to show cell internal resistance as a series-connected external resistor.

When a cell or battery is on charge, the applied terminal voltage must be greater than the EMF so that current is forced against the opposition of the EMF. Since the effective voltage is the difference between the terminal voltage and the EMF,

$$U = E + IR_c$$

$$I = U - \frac{E}{R_c}$$

from which we get

$$R_c = \frac{U - E}{I} \text{ and } U = E + IR_c$$

Compare this with $U = E - IR_c$ for a discharging cell.

Example 17.3
A cell with an EMF of 2 V and an internal resistance of 0.08 Ω is to be charged at 5 A. What terminal voltage must be applied?

$$U = E + IR_c = 2 + (5 \times 0.08) = 2 + 0.4 = 2.4 \text{ V}$$

Example 17.4
A cell is charged at 10 A when a terminal voltage of 2.7 V is applied. If the cell EMF is 2.2 V, what is the internal resistance?

$$R_c = \frac{U - E}{I} = \frac{2.7 - 2.2}{10} = 0.05 \text{ Ω}$$

17.6   Batteries

A single cell is often incapable of providing a high enough voltage for practical purposes, so several cells are connected in series to form a battery. Figure 17.12 shows six cells connected in this way. The total EMF of a battery of cells connected in series is given by multiplying the number of cells by the EMF of each cell, and

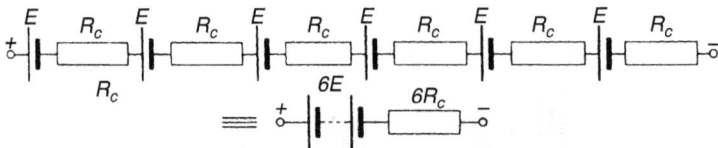

Figure 17.12   Cells connected in series

its internal resistance by multiplying the number of cells by the internal resistance of each cell.

Example 17.5

(a) A cell of internal resistance 0.05 Ω and EMF 2.2 V is connected to a 0.95 Ω resistor. What current will flow?
(b) What current will flow if the same resistor is connected to a battery of six series-connected cells of this specification?

(a) *One Cell*

$$I = \frac{E}{R + R_c} = \frac{2.2}{0.95 + 0.05} = \frac{2.2}{1} = 2.2 \text{ A}$$

(b) *Six Cells*

Battery EMF $= 6 \times 0.05 = 0.3$ Ω

$$I = \frac{E}{R + R_c} = \frac{13.2}{0.95 + 0.3} = 10.6 \text{ A}$$

Note that the use of a battery of six cells has not increased the current six times.

Cells are connected in parallel when the current provided by each cell will be smaller than that required for an acceptable fall in terminal voltage. It is important to notice that cells must never be parallel-connected unless they are identical in terms of EMF and internal resistance. Discrepancies in EMF will result in internal circulating currents in the battery or unequal load-sharing. Both these faults cause rapid deterioration of healthy cells. The EMF of a parallel-connected battery is that of each cell. Its internal resistance is the resultant of the individual internal resistances connected in parallel. The arrangement is shown in Figure 17.13.

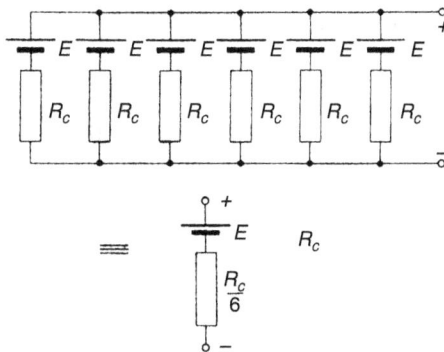

Figure 17.13   Cells connected in parallel

Example 17.6

A cell of EMF 1.6 V and with an internal resistance of 0.3 Ω is connected to a 0.1 Ω resistor.

(a) What current flows?
(b) Find the current if six of these cells are connected in parallel to the same load.

(a) ***Cell current***

$$I = \frac{E}{R + R_c} = \frac{1.6}{0.1 + 0.3} = 4 \text{ A}$$

(b) ***Battery current***

$$E = 1.6 \text{ V}$$

Internal resistance $R_c = \dfrac{0.3}{6} = 0.05 \ \Omega$

$$I = \frac{E}{R + R_c} = \frac{1.6}{0.1 \ + \ 0.05} = 10.7 \text{ A}$$

Series–parallel arrangements of identical cells are used where high voltage and high current are necessary (see Figure 17.14). EMF will be found by:

EMF per cell × The number of cells connected in a series

Overall internal resistance will be given by the following expression:

$$\frac{\text{Resistance of each cell} \times \text{Number of cells in series}}{\text{Number of parallel groups}}$$

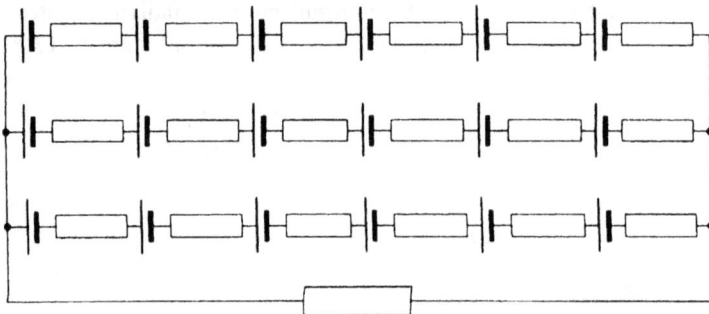

Figure 17.14   Cells connected to load in series-parallel

Example 17.7

Eighteen cells, each of EMF 2.4 V and internal resistance 0.05 Ω are connected in three banks of six cells in series. The three banks are then connected in parallel with each other and with a resistor of 1.9 Ω. Find the current flow in the resistor The arrangement is shown in Figure 17.14.

Battery EMF $= 6 \times 2.4$ V $= 14.4V$

Battery internal resistance $= \dfrac{0.05 \times 6}{3} = 0.1\ \Omega$

$I = \dfrac{E}{R + R_c} = \dfrac{14.4}{1.9 + 0.1} = 7.1$ A

17.7 Capacity and efficiency

A battery of cells is a device for storing energy. The energy stored is known as the capacity of the battery. This is usually measured in ampere hours (Ah). Energy is given by:

$w = VIt$ or $w = Pt$

This method of measurement is not strictly correct. If, however, a battery is capable of providing a current of 5 A for 10 h, it is said to have a capacity of 50 Ah. Such a battery could not be expected to provide current of 10 A for 5 h, since capacity decreases as the current taken increases. Capacity is therefore based on a definite discharge time, usually 10 h when the capacity is quoted at the 10 h rate.

Losses occur both in converting electrical energy to chemical energy (charging) and in the reverse operation of discharging. No cell can be 100% efficient. The actual efficiency is given, as usual, by

Efficiency $= \dfrac{\text{Output}}{\text{Input}} \times 100\%$

There are two methods of measuring the efficiency of a cell.

Ampere $-$ hour efficiency $= \dfrac{\text{Output (discharge)}}{\text{Input (charge)}} \times 100\%$

A cell in good condition is likely to have an ampere-hour efficiency in the region of 80%, which is high since it assumes that charging and discharging terminal voltage are the same. In fact, charging terminal voltage is the sum of cell EMF and internal voltage drop, whereas discharging terminal-voltage drop is given by their difference.

A theoretically truer method of calculating efficiency is the watt-hour method:

$$\text{watt-hour efficiency} = \frac{\text{Average discharge, watt hours}}{\text{Average charge, watt hours}} \times 100\%$$

Watt-hour efficiency for a cell in good condition is likely to be in the region of 65%.

Example 17.8

A discharged 12 V battery is charged for 10 h at 12 A, the average charging terminal voltage being 15 V. When connected to a load, a current of 10 A for 9 h at an average terminal voltage of 12 V discharges the battery. Calculate:

(a) the ampere-hour efficiency
(b) the watt-hour efficiency.

(a) *The ampere-hour efficiency*

$$\text{Ampere-hour efficiency} = \frac{10 \times 9}{10 \times 12} \times 100\% = 75\%$$

(b) *The watt-hour efficiency*

$$\text{Watt-hour efficiency} = \frac{10 \times 9 \times 12}{10 \times 12 \times 15} \times 100\% = 60\%$$

17.8   Summary of formulas for Chapter 7

Discharging

$$U = E - IR_c \quad E = U + IR_c \quad R_c = \frac{E - U}{I}$$

Charging

$$U = E + IR_c \quad E = U - IR_c \quad R_c = \frac{U - E}{I}$$

where U = cell or battery terminal voltage, V; E = cell or battery EMF, V; R_c = cell or battery internal resistance, Ω; and I = charge or discharge current, A.

Internal resistance

Series

For n identical cells in series, internal resistance = nR_c
where R_c = internal resistance of each cell, Ω
EMF = nEE = EMF of each cell.

Parallel

For m identical cells in parallel,

Internal resistance $= \dfrac{R_c}{m}$

EMF $= E$

For a group of m parallel sets, each of n cells in series, z

Internal resistance $= \dfrac{nR_c}{m}$

EMF $= nE$

17.9 Exercises

1. Sketch one of each of any form of (a) primary cell and (b) secondary cell. Label their component parts clearly. Describe how each cell operates, and the EMF in each case.

2. Describe a nickel–cadmium (alkaline cell). (a) Give the characteristic charge and discharge curves, and discuss briefly the advantages and disadvantages of this form of secondary cell. (b) Make a sketch of any one form of primary cell, labelling the separate parts.

3. Make a labelled sketch showing a section through a single-cell dry battery commonly used in hand torches.

4. A cell of EMF 1.5 V has an internal resistance of 0.2 Ω Calculate its terminal PD if it delivers a current of 0.5 A.

5. A cell of EMF 2.2 V and internal resistance 0.05 Ω has a 1.05 Ω resistor connected across its terminals. Calculate the current flow and the terminal PD of the cell.

6. The EMF of a cell is measured with a high-resistance voltmeter on open circuit, and is found to be 1.45 V. When a current of 1 A is drawn from the cell, the terminal PD falls to 1.25 V. What is the internal resistance of the cell?

7. Describe a lead–acid secondary cell. Explain briefly the changes in the cell during charge and discharge. The potential difference between the terminals of a lead–acid cell on open circuit was 2.18 V; when the cell was discharging at the rate of 9 A the terminal PD was 2.02 V. Calculate the internal resistance of the cell.

8. Describe, with sketches, a lead–acid secondary cell and state briefly the chemical changes in the cell during charge and discharge. (The chemical formulas are not required.) Explain the importance of a low internal resistance. A lead–acid cell discharging at the rate of 6 A has a terminal PD of 1.95 V. On open circuit, the PD is 2.1 V. Calculate the internal resistance of the cell.

9.   The electrolyte in a lead–acid cell is ....
10. A battery consists of six 2 V cells in series. Calculate the EMF of the battery.
11. Six cells, each of EMF 1.5 V and internal resistance 0.2 Ω, are connected in series to a 1.8 Ω resistor. Calculate the current delivered, and the battery terminal voltage on load.
12. What is the total EMF when a number of cells are connected in series?
13. What is the purpose of connecting a number of cells in parallel?
14. A lead–acid battery for an electric truck has 15 series-connected cells, each with an EMF of 2.3 V and an internal resistance of 0.01 Ω. Calculate the terminal voltage when the battery delivers a current of 50 A.
15. A cell has an open-circuit terminal voltage of 2.1 V and an internal resistance of 0.1 Ω. Calculate the terminal PD of the cell on charge when the charging current Is:
     (a) 2 A
     (b) 10 A.

16. State the difference between a primary and a secondary cell.
17. A battery comprises five primary cells, each cell having an EMF of 1.1 V and a rated current of 2 A. Calculate the EMF and the rated current of the battery when the cells are connected
     (i) in series
     (ii) in parallel.

     Draw circuit diagrams for each
18. A battery is made up of 12 identical cells connected in series. Each cell has an EMF of 2 V and an internal resistance of 0.05 Ω. What terminal voltage must be applied to the battery if a charging current of 20 A is required?
19. A battery with an EMF of 12 V charges at 10 A when 16 V is applied to it. What is the internal resistance of the battery?
20. A cell with an internal resistance of 0.15 Ω has a terminal PD of 1.8 V when charging at 5 A. What is the EMF of the cell?
21. A lead–acid battery comprises 50 cells in a series, each of open-circuit EMF 2 V and internal resistance 0.02 Ω. Calculate the terminal voltage
     (a) when supplying a load of 10 A
     (b) when being charged at 10 A.

22. Three cells, each of EMF of 1.4 V and internal resistance 0.3 Ω, are connected in parallel to a 0.9 Ω resistor. Calculate the current in the resistor and the battery terminal PD.
23. Twelve lead–acid cells, each of EMF 2.1 V and internal resistance 0.015 Ω, are connected in three series banks of four cells. The banks are connected in parallel to a load resistor. If a current of 20 A flows in this resistor, calculate
     (a) the resistor value
     (b) the terminal PD of the battery.

24. A battery of nine primary cells is connected
    (a) all cells in series
    (b) all cells in parallel
    (c) three sets in parallel, each set consisting of three cells in series.

    Each cell has an EMF of 1.4 V and an internal resistance of 0.45 Ω. The battery terminals are connected to a circuit of resistance 7.2 Ω. Calculate in each case
    (i) the current in the 7.2 Ω resistance
    (ii) the voltage drop across the resistance.

25. A discharged lead–acid battery is charged at 5 A for 15 h at an average voltage of 7.2 V. On discharge, the battery gives 6 A for 10 h at an average PD of 6 V.

26. Calculate the ampere-hour and watt-hour efficiencies

17.10 Multiple-choice exercises

17M1   A secondary cell is on which
    (a) is less efficient than a primary cell
    (b) cannot be recharged when exhausted
    (c) can be recharged after use by means of a charging current
    (d) can be used only for standby purposes

17M2   The standard form of dry cell or battery is called the
    (a) wet Leclanché type
    (b) zinc chloride type
    (c) lead–acid type
    (d) lithium type

17M3   The positive electrode of a zinc-chloride cell consists of
    (a) a carbon rod
    (b) a zinc case
    (c) manganese dioxide
    (d) plaster of Paris

17M4   The terminal voltage of an alkaline cell is
    (a) 2 V
    (b) 1.25 V
    (c) 9 V
    (d) 1.5 V

17M5   The primary cell which is likely to be chosen for use in widely varying temperature conditions is the
    (a) lead–acid cell
    (b) alkaline cell
    (c) PP9 battery
    (d) Leclanché cell

17M6 A battery for use in an emergency-lighting system is likely to consist of
(a) 12 V car batteries connected in series
(b) lithium cells mounted in wooden boxes
(c) alkaline cells contained in plastic enclosures
(d) lead–acid cells contained in glass or plastic containers

17M7 The capacity of a cell or a battery is measured in
(a) ampere hours
(b) litres
(c) volts
(d) watts

17M8 The specific gravity of the electrolyte in a lead–acid cell is measured using a
(a) voltmeter
(b) measuring jug
(c) ammeter
(d) hydrometer

17M9 The following is NOT one of the advantages of the alkaline secondary dry cell
(a) a much more uniform voltage than zine-chloride types
(b) a very long shelf life
(c) being less expensive than the zinc-chloride type
(d) very good leakage protection

17M10 Naked flames, smoking or sparking contacts must be avoided in a lead–acid battery charging room because
(a) the leads connecting the cells might catch fire
(b) the eyes of those working there will be affected
(c) the gases given off during charging can be explosive
(d) the is a standard fire-prevention drill

17M11 A 2 V cell with a terminal voltage of 1.95 V when delivering a current of 2.5 A has an internal resistance of
(a) 0.02 Ω
(b) 0.2 Ω
(c) 0.78 Ω
(d) 0.8 Ω

17M12 A 12 V battery consists of six 2 V cells each of internal resistance 0.006 Ω. The terminal voltage of the battery when delivering a current of 50 A will be
(a) 11.7 V
(b) 10.2 V
(c) 10 V
(d) 13.8 V

17M13 An emergency-lighting battery consists of fifty 2 V cells, each with an internal resistance of 0.01 Ω. If the cells are connected in series, the total voltage and internal resistance of the battery will be

(a) 100 V and 0.01 Ω

(b) 2 V and 0.5 Ω

(c) 100 V and 0.0002 Ω

(d) 100 V and 0.5 Ω

17M14 Forty-eight lead–acid cells, each of terminal voltage 2.1 V and internal resistance 0.05 Ω, are connected as a series–parallel battery of three series strings of 16 cells connected in parallel with each other. The terminal voltage and internal resistance of the battery will be

(a) 33.6 V and 0.267 Ω

(b) 100.8 V and 2.4 Ω

(c) 6.3 V and 0.267 Ω V

(d) 33.6 V and 0.05 Ω

Chapter 18

Introduction to electronics

18.1 Introduction

An electronic circuit differs from a mains circuit because:

- It is usually extra low voltage.
- The components are solid-state semiconductors which work on different principles to conventional electrical equipment.

They are mostly used for control and monitoring. Most domestic appliances, for example washing machines, microwave ovens, and hobs, are fitted with electronic controls. Information and settings are displayed on a small screen. and in many cases, operation is by touch rather than manual knobs or dials. Video, audio and data signals are generated and processed by electronic circuitry, and close speed control for electric motors can also be achieved using purely electronic systems such as soft start.

18.2 Resistors for electronic circuits

Most of the resistors considered in Chapter 2 are types used in power circuits and must be capable of carrying heavy currents. In electronic circuits, we are concerned typically with currents of the order of milliamperes. Since the currents are low and because large numbers of resistors are needed, it becomes economical to use small resistors whose value is not so accurately known. For example, a 120 Ω marked 10% resistor may have a value anywhere between 108 Ω and 132 Ω. This variation in values is called the tolerance and is usually given as a percentage (e.g. ±10%).

Example 18.1
What are the maximum and minimum acceptable values for a resistor marked at 15 k if its tolerance is

(a) ±20%
(b) ±10%
(c) ±5%

(a) ±20%

> 20% of 15,000 is 0.2 × 15,000 = 3000 Ω or 3 kΩ
> Maximum value: 15 k + 3 k = 18 kΩ
> Minimum value: 15 k − 3 k = 12 kΩ

(b) ±15%

> 10% of 15,000 is 0.1 × 15,000 = 1500 Ω or 1.5 kΩ
> Maximum value is thus 15 k + 1.5 k = 16.5 kΩ
> Minimum value is thus 15 k − 1.5 k = 13.5 kΩ

(c) ±5%

> 5% of 15,000 = 0.05 × 15,000 = 750 Ω
> Maximum value is thus 15,000 + 750 = 15.75 kΩ
> Minimum value is thus 15,000 − 750 = 14.25 kΩ

A very important property of a resistor is its power rating. This is the power that may be dissipated in the resistor continuously without it becoming overheated and depends on the current carried:

$$P = I^2 R \quad I^2 = \frac{P}{R} \quad I = \sqrt{\frac{P}{R}}$$

where I = maximum sustained permissible current, A; P = power rating of resistor, W; and R = resistance of resistor, Ω

Example 18.2
Calculate the maximum permissible current in a 1 kΩ resistor if it is rated at

(a) 0.5 W
(b) 1 W
(c) 2 W

(a) 0.5 W

$$I^2 = \frac{P}{R} = \frac{0.5}{1000} \quad I^2 = 0.0005 \text{ A} \quad I = \sqrt{0.0005} = 0.022 \text{ A or 22 mA}$$

(b) 1 W

$$I^2 = \frac{P}{R} \quad \frac{1}{1000} \quad I^2 = 0.001 \text{ A} \quad I = \sqrt{0.001} = 0.032 \text{ A or 32 mA}$$

(c) 2 W

$$I^2 = \frac{P}{R} = \frac{2}{1000} \quad I^2 = 0.002 \text{ A} \quad I = \sqrt{0.002} = 0.045 \text{ A or 45 mA}$$

Example 18.3

What rating should be chosen for a 12 kΩ resistor which is to carry a current of 9 mA?

$$P = I^2R \quad P = 0.009^2 \times 12,000 = 0.000081 \times 12,000 = 0.972 \text{ W}$$

In practice, the nearest rating above the calculated value would be chosen: 1 W.

The resistor is often too small to display its value and power rating on the component, so a code is used to show its resistance. Two codes are commonly in use:

- Colour
- Digital

Colour code

A series of coloured bands are printed on the resistor. Each colour represents a number, or in some cases a tolerance. The colours and the values they represent are:

- black 0
- brown 1
- red 2
- orange 3
- yellow 4
- green 5
- blue 6
- violet 7
- grey 8
- white 9

The colours are applied in four bands (Figure 18.1). The first band indicates the first figure of the value, the second band the second figure and the third band the number of zeros to be added. The third band is sometimes coloured gold or silver, indicating one-tenth or one-hundredth, respectively, of the first two bands.

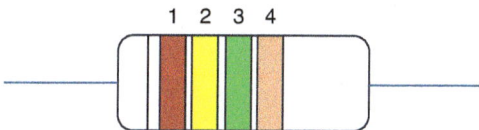

Values	Totals
1: Brown = 1 first figure of value	1 Ω
2: Yellow = 4 second figure of value	14 Ω
3: Green = 5 number of zeros to be added	14,00000 Ω
4: Pink = ±20% tolerance	+168,000 Ω
	−1120000 Ω

Figure 18.1 Example of resistor colour coding

If the bands 1, 2, 3 and 4 are, respectively, red–red–brown–silver, it indicates 220 ±10%. Sometimes, a resistor is specially made to ensure that its resistance does not change as it ages. These 'high-stability' resistors are indicated by a salmon-pink fifth band.

Digital code

This code takes the form of numbers and letters printed on the resistor and is most often applied to wire-wound types.

R indicates a decimal point, so that:

- 1R0 means 1 Ω
- 4R7 means 4.7 Ω
- 68R means 68 Ω and so on

K has a similar function, but indicates values in thousands of ohms, or kilohms, so that:

- 1K0 means 1 kΩ
- 4K7 means 4.7 kΩ
- 82K means 82 kΩ and so on

M serves similar purpose, but indicates values in millions of ohms, or megohms.

- 1M2 means 1.2 MΩ
- 15M means 15 MΩ and so on

With this code, tolerances are indicated by a code letter placed after the value.
The tolerance code is

- B ± 0.1%
- C ± 0.25%
- D ± 0.5%
- F ± 1%
- G ± 2%
- J ± 5%
- K ± 10%
- M ± 20%
- N ± 30%

Examples of the use of this code are

- 4R7J = 4.7 ± 5%
- 6K8F = 6.8 k ± 1%
- 68KK = 68 k ± 10%
- 4M7M = 4.7 M ± 20%

Types of resistor

Many types of resistor are used in electronic circuits. Carbon-composition resistors are most common and are moulded from carbon clay compound into a cylindrical

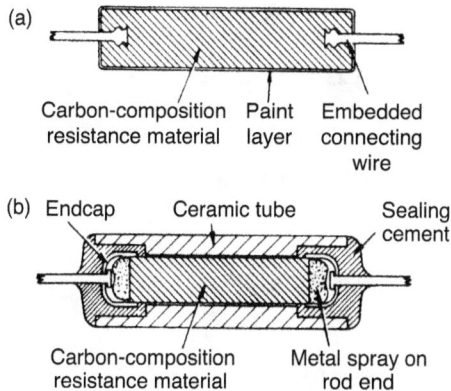

Figure 18.2   Carbon-composition resistors. (a) Uninsulated type; (b) insulated type

shape. The connection is via wire ends. They are made in various sizes with power ratings from 1/8 Ω to 2 Ω (Figure 18.2).

Carbon resistors

Carbon resistors often have higher stability than moulded-carbon types. The resistive film or coating is deposited on a glass tube, which is buried in a plastic moulding. The connecting wires are carried into the ends of the glass tube to conduct heat away from the resistor.

Cracked-carbon film, or pyrolytic resistors

This type has a film of cracked carbon deposited on a ceramic rod. The film is then cut through in a spiral pattern, producing what is effectively a long, thin resistor element wound round the rod. Endcaps with connecting leads and a coating of silicon lacquer complete the construction. Cracked carbon resistors have higher stability than other types.

Metal-film and metal-oxide resistors

These have a similar construction to the cracked carbon film type, but the film, and hence the spiral track, is formed of nickel–chromium or a metal oxide. These resistors are capable of operation at very high temperatures and are accurate with a tolerance of only ±1%.

Wire-wound resistors

Wire-wound resistors are used when the power to be dissipated is high. The resistance wire (nickel–chromium) is usually wound on a ceramic tube and given a vitreous-enamel coating for insulation and protection (Figure 18.3).

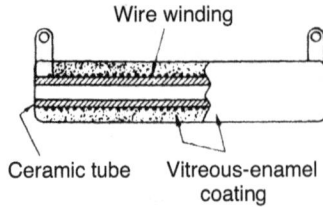

Figure 18.3   Wire-wound vitreous-enamelled resistor

Variable resistors

Usually operated by turning a shaft or moving a slider, and are of two types:

- The wire-wound type (see Figure 4.3 in Chapter 4 for a much higher-power type) is used for lower resistances (up to about 100 kΩ) where higher powers are dissipated.
- The carbon-track type is used for very high resistances (up to about 2 MΩ).

Non-linear resistor or varistor

The resistance of a varistor is not uniform and it does not obey Ohm's Law. Their *VI* characteristics curve is non-linear. One of their applications is in voltage surge protection. When the voltage at the varistor is below its threshold, the resistance is high, which limits the current flow. It is, in effect, an open switch. When the voltage increases, the resistance drops, which allows a large current to flow. The varistor absorbs the current so that it cannot damage the equipment the varistor is protecting. Due to the inverse proportionality of voltage and current, the high current reduces the voltage to safe levels.

Varistors are manufactured from materials such as:

- Silicon carbides
- Zinc oxide

There are three types of varistors:

- Silicon carbide disc type varistor
- Silicon carbide rod type varistor
- Zinc oxide type varistor

Light-dependent resistor (LDR)

The resistance of an LDR depends on the intensity of light to which it is exposed. LDRs are constructed from cadmium sulphide. When it is not illuminated, it boasts only a small number of electrons. When exposed to light, its electrons are ejected. As a result, the LDR's conductivity increases. LDRs are used for photometers (light meters).

Magneto resistor

Magneto resistors are used in electronic compasses and for measuring magnetic field direction and strength. The magneto resistor makes use of the fact that changes in a magnetic field also change the resistance in a conductor associated with that magnetic field.

Permalloy, an alloy made from nickel and iron, is a favoured material for these components. Gold or aluminium bands are wound around the resistor, at an angle of 45° to increase its sensitivity.

18.3   Semiconductor diodes

Diodes are constructed from a crystal of the basic semiconductor material such as:

- Germanium
- Silicon
- Gallium arsenide
- Gallium nitride
- Graphine

The semiconductor has opposite sides treated in such a way that the two halves have different characteristics.

- One-half of the crystal becomes a *p*-type material, with a shortage of mobile electrons.
- The other half becomes an *n*-type material, which has surplus of mobile electrons.

Some of these surplus electrons cross the boundary from the *n*-type half to the *p*-type half of the crystal, making the *p*-type material negatively charged and the *n*-type material positively charged (Figure 18.4).

Electrons can move more easily from the *n*-type region to the *p*-type, but not in the reverse direction. The device behaves as a rectifier. The circuit symbol for the semiconductor diode is shown in Figure 18.5. The symbol's arrow points the direction in which conventional current can flow.

Figure 18.6 shows a typical rectifying circuit and indicates how current can pass through the load during the positive half-cycles of an alternating supply, but not during the negative half-cycles.

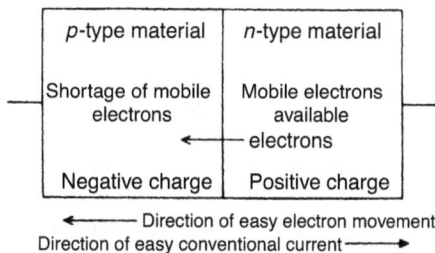

p-type material	*n*-type material
Shortage of mobile electrons	Mobile electrons available
	←——— electrons
Negative charge	Positive charge

←——— Direction of easy electron movement
Direction of easy conventional current ——➤

Figure 18.4   Simple representation of a p–n junction (semiconductor diode)

Direction of easy electron movement

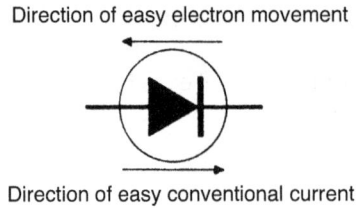

Direction of easy conventional current

Figure 18.5   Circuit symbol for semiconductor diode

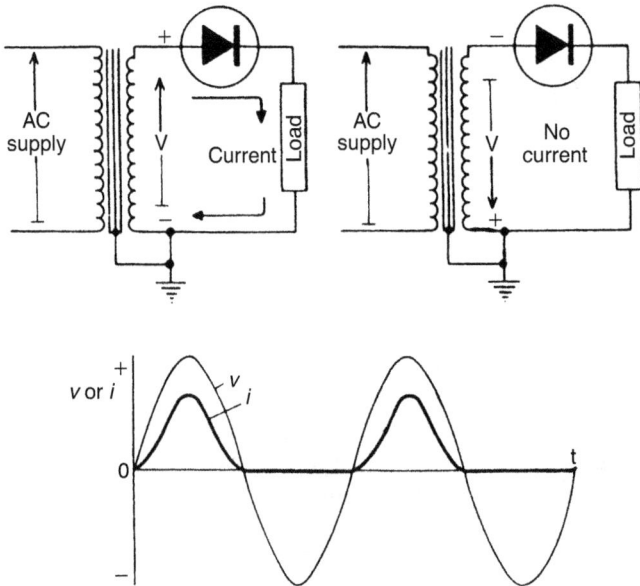

Figure 18.6   Circuit and wave diagrams for semiconductor diode connected to load as halfwave rectifier

In the direction in which current flows easily (the forward direction), there will be no current through the diode until a certain applied voltage is reached. This is the forward volt drop that is approximately:

- 0.3 V for germanium diodes
- 0.7 V for silicon types.

In the reverse direction, the diode will not allow current to flow at all (other than a very small leakage current) unless the applied voltage is large enough to break down the *p–n* junction. When this happens, the diode is destroyed as a rectifier. The reverse-breakdown voltage varies with the construction of the diode, but is seldom less than 25 V for germanium and 75 V for silicon types.

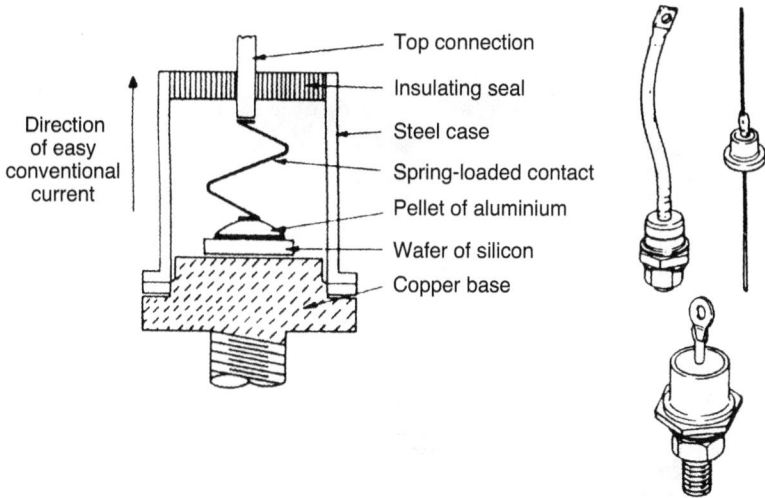

Figure 18.7 Silicon-diode construction

Semiconductor diodes are used widely in industry, their main application being the conversion of an AC supply to DC. Although the action of a single diode does give a direct current, but a series of isolated current pulses, and is quite unlike the output from a DC generator or a battery. By using two or more diodes, the output may be improved, and the addition of capacitors allows a smooth supply to be obtained.

Silicon diodes have a very high current rating indeed and an extremely high reverse breakdown voltage. A silicon diode, no larger than a pea, can carry a current of 15 A, and one 50 mm in diameter and 40 mm long can handle over 600 A. Figure 18.7 shows a typical construction for a silicon diode.

These rectifiers must be kept cool to prevent a breakdown and are often mounted on aluminium castings called heat sinks to increase current rating. Very heavily loaded silicon diodes are sometimes water-cooled.

Four diodes are required to achieve full-wave rectification (see Figure 18.8).

18.4 Transistors

The diode is known as a passive device because under normal circumstances, it is not possible to change its effective operation. The transistor, however, is an active device because its operation can be controlled by the application of a current or a voltage. There are many types of transistors, but we will limit our explanation to just two of the most common types.

Bipolar junction transistor (BJT)

This device is still probably the most common transistor, especially when used in amplifiers. It is composed of three layers of semiconducting material, either:

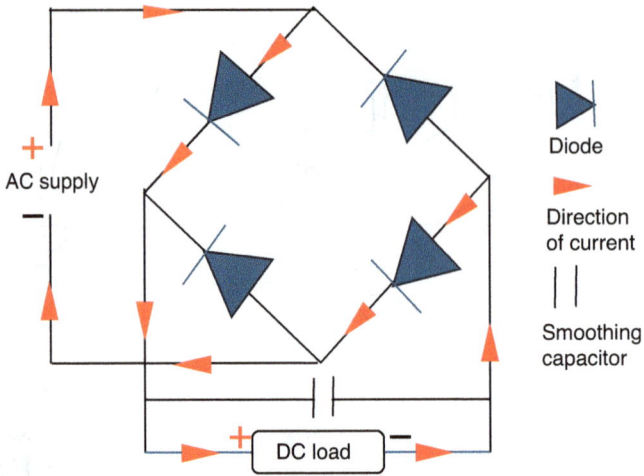

Current flow during positive half of AC sine wave

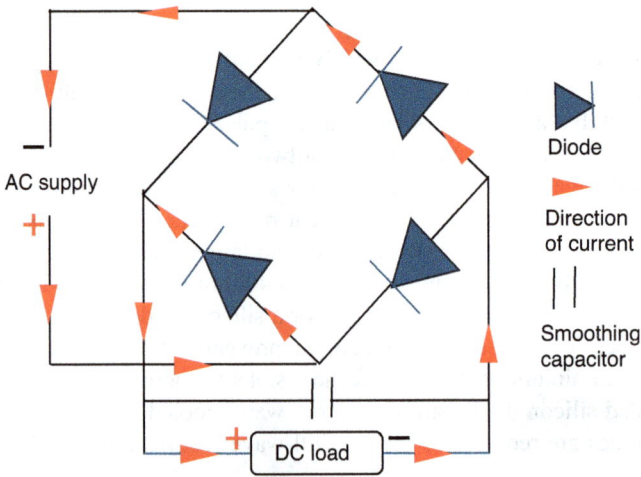

Current flow during negative half of AC sine wave

Figure 18.8   Full wave rectifier principles for both positive and negative halves of an AC sine wave

• A very thin layer of *p*-type sandwiched between two thicker layers of *n*-type – the *npn* transistor
• Two thick layers of *p*-type separated by a thin layer of *n*-type – the *pnp* transistor (now uncommon)

An explanation of *p*-type and *n*-type semiconductor materials was given in Section 18.3. In both cases, the thin central layer is called the base, whilst the two

(a)

(b)

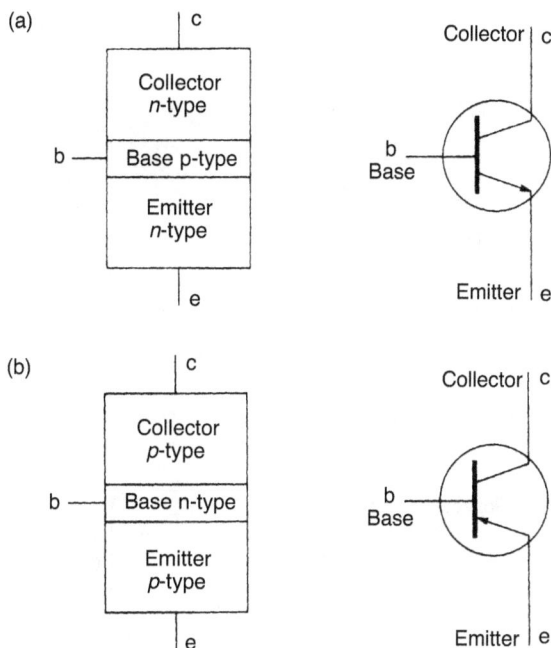

Figure 18.9 npn and pnp types of bipolar junction transistor. (a) npn-transistor arrangement and circuit symbol; (b) pnp-transistor arrangement and circuit symbol.

thicker outer layers are known as the emitter and the collector. The physical arrangements of the two types, together with their circuit symbols, are shown in Figure 18.9.

The operation of the BJT transistor is complicated, but a simplified explanation of how the *npn*-type works is as follows. First, we must remind ourselves of the operation of the diode, because the BJT transistor is in some respects like two diodes connected together.

- With the diode, we can consider that in the forward direction, in which current flows more easily, the device will have low resistance.
- When the applied voltage is such that the diode behaves in this way, it is said to be forward-biased and in this condition its resistance is comparatively low.
- If the applied voltage charge is reversed, the diode is said to be reverse biased, in which condition it has very high resistance and hardly any current flows.

The BJT transistor has two junctions:

- that between the emitter and base (the *e–b* junction)
- that between the collector and base (the *c–b* junction)

The transistor is connected in a circuit as shown in Figure 18.10, and the applied voltages cause the *e–b* junction to be forward biased so that its resistance is low and the *c–b* junction reverse biased, resulting in very high resistance.

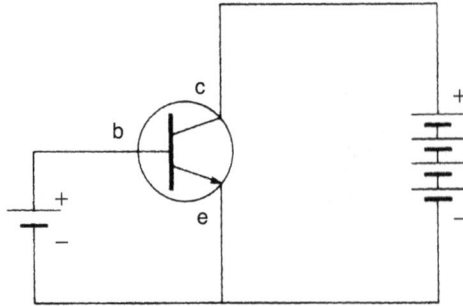

Figure 18.10   Biasing voltages for the npn bipolar junction transistor. Emitter-base (e-b) junction forward biased (low resistance). Collector-base (c-b) junction reverse biased (high resistance).

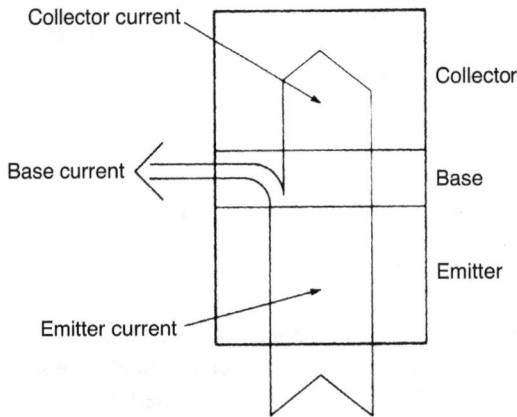

Figure 18.11   Explanation of amplification operation

The amplifying action of the transistor may be simply explained by reference to Figure 18.11. Current (in the form of a flow of electrons) enters the base region from the emitter, driven by the emitter-base voltage against the low resistance of the *e–b* junction. Because the resistance is low, a small change in voltage will result in a large current change.

As the base region is very thin, most of the electrons entering from the emitter are unable to leave by the base connection but 'blunder' across the high-resistance collector–base (c–b) junction. Thus, a small voltage change in the emitter–base low-resistance circuit results in a significant current change in the collector–base high-resistance circuit.

As the *c–b* junction has such high resistance, the current flow is hardly affected by the connection of a high-value resistor (usually 10–20 k) in the collector circuit (the collector load). Thus, a small voltage change in the emitter–base circuit results

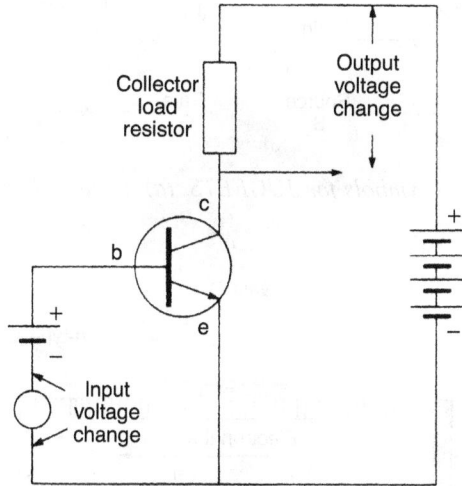

Figure 18.12   Transistor as a voltage amplifier

in a significant current change in the collector load resistor and hence a much higher change in the voltage across it. Hence, the circuit has performed as a voltage amplifier (see Figure 18.12).

It must be appreciated that this description is grossly oversimplified, but hopefully will give some inkling of how the device performs its function. The BJT can be considered as a resistor whose resistance is reduced as the base current increases and increases as the base current becomes smaller. It is called an active device because the circuit resistance is controlled by electrical means rather than by physical movement, such as the operation of a slider.

Field-effect transistor (FET)

This type of transistor is used very widely indeed in data and logic circuits, such as those in computers, programmable controllers and so on. Like the BJT, the FET is an active device whose resistance is controlled by the application of a variable low voltage to the gate connection. The two most important types of FET are the:

- Junction-gate type (JFET)
- Insulated-gate type (IGFET)

JFET

This transistor can be made as *n*-channel or *p*-channel types, the former being more common. Circuit symbols for the two types are shown in Figure 18.13.

The *n*-channel type consists of a cylinder of *n*-type silicon surrounded by a *p*-type shroud or tube, so that the conducting channel between the end contacts is through the *n*-type material. The general arrangement is shown in Figure 18.14, which also makes it clear that the connections to the ends of the channel are called

(a)

Drain
D

Gate
G

Source
S

(b)

Drain
D

Gate
G

Source
S

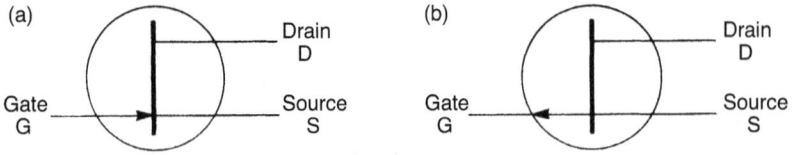

Figure 18.13   Circuit symbols for JUGFETS. (a) n-channel type; (b) p-channel type.

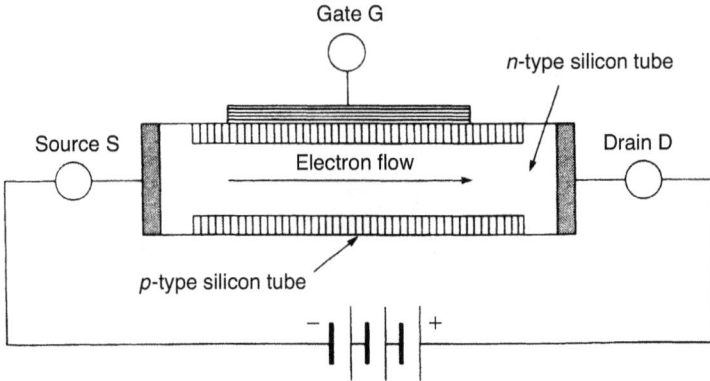

Gate G

n-type silicon tube

Source S

Electron flow

Drain D

p-type silicon tube

− +

Figure 18.14   Arrangement of an n-channel JUGFET

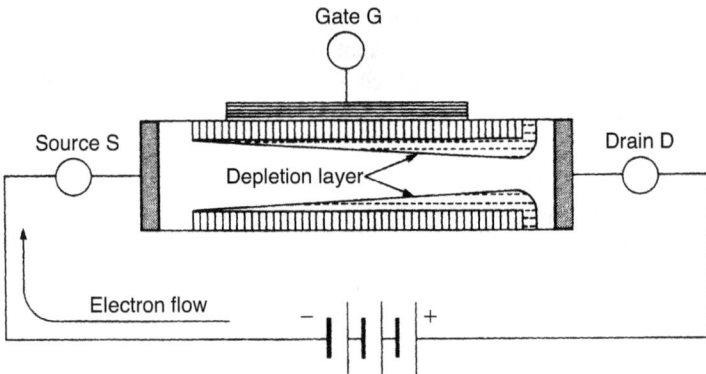

Gate G

Source S

Depletion layer

Drain D

Electron flow

− +

Figure 18.15   Negative gate potential reduces the cross-sectional area of the current path

the source (S) and the drain (D), while the connection to the surrounding cylinder is called the gate (G).

If the gate is made negative, a depletion later is formed within the silicon tube, as shown in Figure 18.15. This reduces the cross-sectional area of the channel available for electron flow and thus reduces the current.

If the negative potential on the gate is increased further, a point will be reached where the depletion layers meet, completely closing the *n*-channel. This is called the pinch-off point and no current flows in the device. This condition is illustrated in Figure 18.16.

The simplified account above makes it clear that the current flowing in the field effect transistor can be controlled by the level of the negative voltage applied to the gate. In most cases, FETs are used at voltages not exceeding 5 V. The *p*-channel-type JUGFET operates in much the same way as the *n*-channel-type, except that the tube is *n*-type and the cylinder is *p*-type.

IGFET

As the name 'insulated-gate field-effect transistor' indicates, this device has its gate connection insulated from the channel by a layer of silicon dioxide. It again exists in *n*-channel and *p*-channel forms. Circuit symbols are shown in Figure 18.17 and a simplified construction in Figure 18.18.

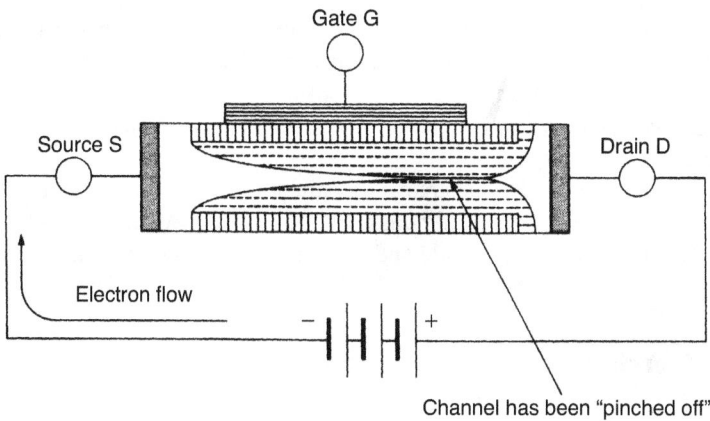

Figure 18.16   Increased negative gate voltage has widened the depletion layers, 'pinching off' the channel

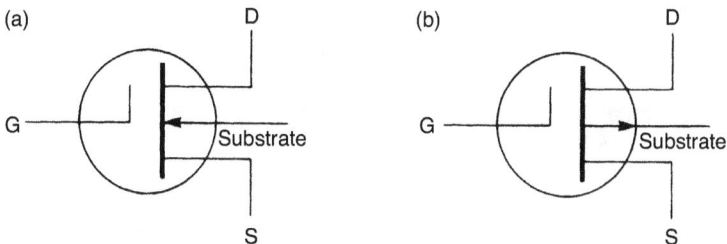

Figure 18.17   Circuit symbols for IGFETS. (a) n-channel type; (b) p-channel type.

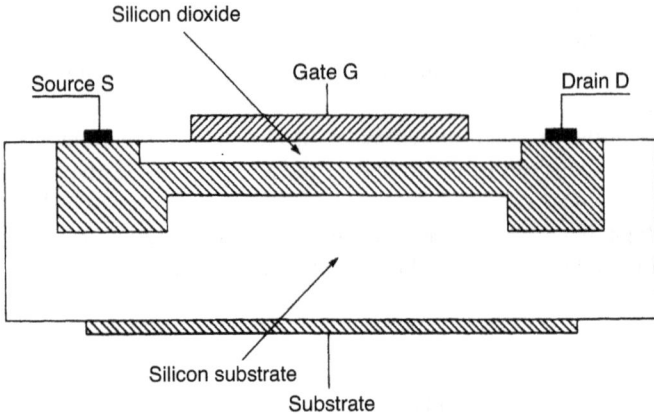

Figure 18.18   Simplified construction of an IGFET

As in the JFET, pinch-off can be achieved by making the insulated gate negative with respect to the source, when the transistor is said to be depleted. However, in this type of transistor, the opposite effect is also possible: a positive voltage on the gate increases the current in the channel when the transistor is said to be enhanced.

The connection to the substrate is available and may be used as the other terminal for biasing the gate, although the substrate is often connected to the source. Because of the metal oxide and silicon used in construction, the IGFET is sometimes referred to as a MOSFET.

18.5   Exercises

1. What are the possible maximum and minimum values of a resistor marked 680 if its tolerance is
    (a) ±20%
    (b) ±10%
    (c) ±5%
    (d) ±2%

2. A resistor is marked as 180 k but measured at 192 k. What tolerance does this indicate?

3. Calculate the maximum currents for the following resistors:
    (a) 2.7 k, 0.5 W
    (b) 68, 0.125W
    (c) 120 k, 1W
    (d) 1.2 M, 0.5W
    (e) 43 k, 2W

4. Give the values and tolerances of the resistors having the following codes:
   (a) orange, orange, yellow, gold
   (b) brown, black, orange, silver
   (c) grey, red, green
   (d) red, violet, gold, silver
   (e) 390RK
   (f) 2K2K
   (g) 3M9M
   (h) 39RG
   (i) 47KJ
   (j) 5K6N

5. Use sketches to assist descriptions of the following types of resistor:
   (a) carbon-composition
   (b) carbon-film
   (c) pyrolytic
   (d) metal-oxide
   (e) wire-wound

6. Draw a circuit diagram to show a semiconductor diode connected in series with a load to an AC supply. Draw a wave diagram of the supply voltage and the circuit current, describing why the current wave has the form shown.

7. Sketch the circuit symbol of a *pnp* bipolar-junction transistor, labelling the connections. With the aid of a sketch showing its construction, explain why a small voltage change across the low-resistance base–emitter region of an *npn* BJT will result in a much larger than expected current change in the high-resistance collector-base region.

8. Draw a circuit diagram of an *npn* bipolar-junction transistor connected for use as a voltage amplifier.

9. Explain in simple terms the construction and the operation of a junction-gate field-effect transistor (JUGFET). In the course of your explanation, indicate the meaning of the terms depletion layer and pinch-off.

10. Draw the circuit symbols for both *n*-channel and *p*-channel insulated-gate field effect transistors (IGFETs), labelling the connections.

18.6 Multiple-choice questions

18M1   A resistor marked as having a value of 47 k ± 10% may have an actual value between
   (a) 42.3 k and47k
   (b) 37 k and57k
   (c) 42.3 k and51.7 k
   (d) 44.7 k and49.4 k

18M2   A suitable power rating for a 68 k resistor carrying a current of 3.5 mA is
(a) 250 W
(b) 0.1 W
(c) 2W
(d) 1W

18M3   A1.2M resistor with a power rating of 0.5 W should carry no more current than
(a) 0.65 mA
(b) 0.42 μA
(c) 0.65 A
(d) zero

18M4   The coloured band printed on a resistor to represent the number 6 is
(a) red
(b) violet
(c) yellow

18M5   The third coloured band on a resistor represents
(a) the power rating of the resistor
(b) the tolerance of the resistor
(c) blue
(d) the number of zeros to be added to the value given by the first two colour bands
(e) the temperature at which the resistor will safely operate

18M6   If the four coloured on a resistor are from left to right, brown, grey, yellow and silver, the resistor value and tolerance will be
(a) 180 k ±10%
(b) 18M ±10%
(c) 27 k ±5%
(d) 18k ±10%

18M7   A56 resistor with a tolerance of ±5% is indicated by bands with colours
(a) orange–blue–grey–silver
(b) green–blue–black–gold
(c) blue–green–black–silver
(d) green–blue–black–salmon pink

18M8   If a resistor is marked 1M8M, its value and tolerance is
(a) 18 M ±10%
(b) 1.8 M ±5%
(c) 1.8 M ±20%
(d) 1.8 M ±20%

18M9   The digital marking of 680 kJ on a resistor indicates a value and tolerance of
  (a) 680 k ±5%
  (b) 6.8 k ±10%
  (c) 68k ±10%
  (d) 68k ±20%

18M10 A resistor of 33k with a tolerance of ±10% will carry the digital marking
  (a) 33KJ
  (b) 3K3K
  (c) 33MK
  (d) 33KK

18M11 The type of resistor to select if the power it is to dissipate will be 15 W
  (a) wire wound
  (b) cracked carbon film
  (c) carbon composition
  (d) metal-oxide

18M12 The arrow in the symbol used for a semiconductor diode points in the direction of
  (a) easy electron movement
  (b) the positive connection
  (c) easy conventional current flow
  (d) the connection to the supply

18M13 A semiconductor material with mobile electrons and an overall positive charge is called
  (a) a *p*-type material
  (b) a rectifier
  (c) an *n*-type material
  (d) a diode

18M14 The diagram below is for a
  (a) power resistor
  (b) half-wave rectifier
  (c) heating circuit
  (d) full-wave rectifier load

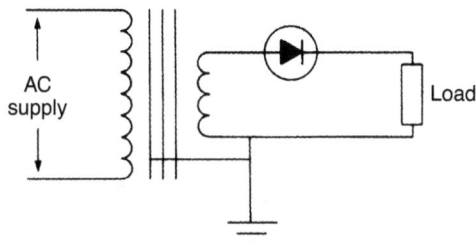

18M15 The most common material used in semiconductor devices is
(a) copper oxide
(b) germanium
(c) selenium
(d) silicon

18M16 The three regions of the bipolar-junction transistor are called the
(a) emitter, base and collector
(b) diode, collector and base
(c) emitter, forward bias and collector
(d) source, gate and drain

18M17 The emitter–base (e–b) junction of a bipolar-junction transistor is
(a) reverse biased to have very high resistance
(b) forward biased to have low resistance
(c) extremely thin
(d) composed of a *p*-type silicon cylinder

18M18 The circuit symbol for an *npn* bipolar transistor is

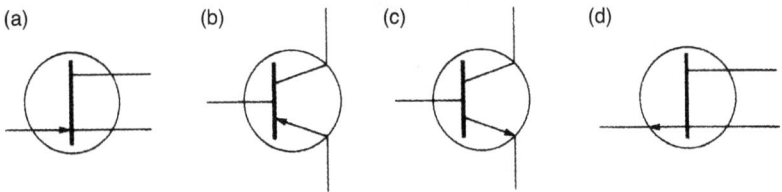

18M19 The three connections of a junction-gate field-effect transistor are called the
(a) emitter, base and collector
(b) source, depletion layer and gate
(c) source, gate and drain
(d) positive and negative

18M20 When the depletion layers in the channel of a field-effect transistor meet, the device is said to be
(a) non-conducting
(b) closed
(c) switched-off
(d) pinched-off

18M21 The circuit symbol shown represents
  (a) a silicon diode
  (b) *ap*-channel-type insulated-gate field-effect transistor
  (c) an *npn* bipolar-junction transistor
  (d) an *n*-channel-type junction-gate field-effect transistor

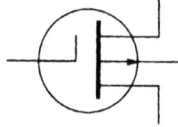

Answers to numerical exercises and multiple choice tests

Chapter 1

1.10 Exercises

2. (a) 0.003264 MΩ
   (b) 0.000034 F
   (c) 560 V
   (d) 60,000 mm^2
3. (a) $I = \frac{P}{V}$
   (b) $L = \frac{X_L}{2\pi F}$
   (c) $F = \frac{1}{Xc2\pi C}$
   (d) $nr = \frac{ns - \%slip}{ns}$
4. 19.44%
5. 14.96
6. 129 m
7. 63.38%
8. 18.12 W
9. (a) $\text{Cosine} = \dfrac{\text{Adjacent}}{\text{Hypotenuse}}$

   (b) $\text{Sine} = \dfrac{\text{Opposite}}{\text{Hypotenuse}}$

   (c) $\text{Tangent} = \dfrac{\text{Opposite}}{\text{Adjacent}}$

10. 16.79°
11. 0.79
12. 7.8 mm
13. 74.12 mm
14. (a) 2.29 m, (b) 1.15 m, (c) 4.16 m^2
16. 1.18:1
17. 9-socket outlets
19. 10.54 Ω
20. 144.25 μF

1.11   Multiple-choice questions

1M1 (c)
1M2 (d)
1M3 (a)
1M4 (a)
1M5 (c)
1M6 (b)
1M7 (d)
1M8 (b)
1M9 (a)
1M10 (c)
1M11 (d)
1M12 (c)

Chapter 2

2.6   Multiple-choice questions

2M1 (b)
2M2 (d)
2M3 (d)
2M4 (a)
2M5 (d)
2M6 (a)
2M7 (b)
2M8 (b)
2M9 (a)
2M10 (b)

Chapter 3

3.15   Exercises

1. 1200 C
2. 100 min
3. 12.5 A
5. 120,000 J
6. 2.89 μV
7. 57.5 A
9. 19.2
10. 15 A
11. 2.875 A, 14.4 V, 43.1 V, 57.5, 115 V
12. 2 Ω
13. 459.8 Ω

14. (a) 1.5 Ω
    (b) 3.6 M
    (c) 80 μΩ
15. (a) overcurrent, heating
    (b) coulomb
    (c) copper, aluminium
    (d) the bell, the motor
    (e) PVC, mica
    (f) negative, positive
    (g) heating
    (h) negative
    (i) 6 Ω
    (j) 5 Ω
    (k) current × time
    (l) 360 mA
    (m) 3300 V
16. 30 V, 3 A, 3 Ω
18. (a) 78 V
    (b) 312 V
19. 6 Ω
20. 7.8 Ω

3.16   Multiple-choice questions

3M1 (b)
3M2 (a)
3M3 (d)
3M4 (d)
3M5 (b)
3M6 (d)
3M7 (a)
3M8 (c)
3M9 (a)
3M10 (d)
3M11 (b)
3M12 (a)

Chapter 4

4.9   Exercises

1. 2.16 Ω
2. 2.5 Ω
3. 0.194 Ω
4. 0.75 Ω

5. 2.5 times
7. 0.342 Ω
8. 3.82 Ω
9. 2500 μΩmm
10. 439 m
11. 109.9 Ω
12. 15.06 Ω
13. 200°C
14. 17.0 Ω
15. (a) 192 V
    (b) 221 V
    (c) 230 V
    (d) 240 V
16. 3 V
17. 226 V
18. 0.383 Ω
19. 4.31 A
20. 195.9 V

4.10   Multiple-choice questions

4M1 (b)
4M2 (c)
4M3 (a)
4M4 (d)
4M5 (a)
4M6 (c)
4M7 (b)
4M8 (d)
4M9 (a)
4M10 (b)
4M11 (d)
4M12 (c)

Chapter 5

5.8   Exercises

1. 19,620 N
2. 51.0 kg
3. 200 N
4. (a) 60,000 N/m^2 (b) 800 N/m^2
5. 66.7 N
6. 20 Nm
7. 4200 J, 35 W

8. 22500 J
9. 4 kW
10. 480 N, 2.4 N/mm^2
11. (a) 600 N, (b) 9
12. 375 N
13. 160 N
14. 0.139 m, 222 Nm
15. 500 mm, 500 Nm
16. 130 teeth, 446 r/min
17. 2800 N, 82°
18. 127° from 6000 N, 10,000 N
19. 600 N
20. (b) 1156 N (c) 1250 N

5.9   Multiple-choice questions

5M1 (c)
5M2 (a)
5M3 (d)
5M4 (a)
5M5 (b)
5M6 (d)
5M7 (c)
5M8 (b)
5M9 (a)
5M10 (d)
5M11 (d)
5M12 (b)
5M13 (a)
5M14 (c)
5M15 (b)
5M16 (c)

Chapter 6

6.3   Exercises

1. 1.44 kW
2. 48 W
3. 2.3 kW
4. 3.26 A
5. 10 A
6. 300 V
7. 1.33 kW
8. 200 kWh or 720 MJ

9. £4.80
10. 744 W
11. 2.7 kW
12. 2.0 kW, 8.70 A
13. 0.37 A, 30.5 A, 10 A, 30 A
14. £1.62
15. 80%
16. 21.6 kW
17. 6 kW, 11.8 kW

6.4　Multiple-choice questions

6M1　(c)
6M2　(a)
6M3　(d)
6M4　(a)
6M5　(b)
6M6　(c)
6M7　(b)
6M8　(d)
6M9　(a)
6M10　(b)

Chapter 7

7.8　Exercises

1. (a) 333 K (b) 198 K (c) 1273 K
2. (a) 47°C (b) 1227°C (c) −33°C
3. 2.68 MJ
4. 3.76 MJ, 1.04 kWh
5. 9.01 kWh
6. 64.1 MJ
7. 16.7 MJ
8. 8370 J/s
9. 1100 kWh
10. 34.9°C
11. 120°C
12. 14.5°C
13. 27 min 55 s
14. 10.5 kW
15. 369°C

7.9   Multiple-choice questions

7M1 (c)
7M2 (a)
7M3 (d)
7M4 (b)
7M5 (b)
7M6 (a)
7M7 (d)
7M8 (b)
7M9 (c)

Chapter 8

8.10   Exercises

4. 1.3 T
5. 6.72 μWb
6. (a) 4800 At/m (b) 4.82 μWb (c) 6.03 mT
7. 2120 At
8. 965 μWb, 1.21 T
9. 0.462 mWb
11. (a) increase (b) increase (c) decrease (d) decrease (e) increase

8.11   Multiple-choice questions

8M1 (b)
8M2 (d)
8M3 (a)
8M4 (b)
8M5 (c)
8M6 (a)
8M7 (d)
8M8 (b)
8M9 (b)
8M10 (d)
8M11 (c)
8M12 (a)
8M13 (b)
8M14 (d)

Chapter 9

9.8   Exercises

1. (a) into paper
   (b) out of paper
   (c) north pole at bottom
   (d) right to left
2. 2 m
3. 3 m/s
4. 2 m/s
5. 0.48 V
6. 4 V, 800 V
7. 4 V
8. 0.167 s
9. 6.67 mWb

9.9   Multiple-choice questions

9M1 (b)
9M2 (d)
9M3 (a)
9M4 (c)
9M5 (a)
9M6 (d)
9M7 (a)
9M8 (d)
9M9 (b)
9M10 (a)
9M11 (a)
9M12 (d)
9M13 (c)

Chapter 10

10.6   Exercises

2. 104 turns
3. 125 V
4. 1992 turns
8. (a) 480 turns (b) 23 V
10. 4.17 A
11. 100 A
12. 2500 turns, 1.5 kV

10.7   Multiple-choice questions

10M1 (c)
10M2 (b)
10M3 (a)
10M4 (c)
10M5 (b)
10M6 (a)

Chapter 11

11.8   Exercises

1. 4.44 A
2. 13.3 m
3. 1.1 T
5. 4000 N
6. 0.25 mA
8. 1.11 T
9. 20 m
10. (b) 12 N

11.9   Multiple-choice questions

11M1 (c)
11M2 (d)
11M3 (a)
11M4 (d)
11M5 (b)
11M6 (c)
11M7 (a)
11M8 (c)
11M9 (b)
11M10 (d)

Chapter 12

12.11   Exercises

1. 1000 Hz
2. 16.7 ms
3. (a) 66.7 Hz (b) 150 V (c) 162 V (d) 1.105
4. 141 V, 127 V
5. 20 ms
6. (a) 25 Hz (b) 35 A (c) 38.7 A (d) 1.105 (e) −40 A
7. 127 V, 141 V, 1.11
8. 9.01 A, 14.1 A
9. 207 V, 325 A

10. 191 A, 212 A
11. (b) 70.7 A
12. 141 A
13. (a) 141 A (b) 135°
14. 24.3 A leading by 17°
15. 100 V, 60°
16. (a) 11.5 A (b) 5.75 A (c) 50 A (d) 10 A (e) 15 A
17. (a) 314.2, 0.732° (b) 7.54, 53.0 A (c) 377, 0.265 A
18. (a) 318, 0.723 A
    (b) 7.96 k, 1.51 mA
    (c) 17.7, 0.17 A
    (d) 3.18 M, 7.54 μA
    (e) 3.98, 0.251 A
19. 125 Ω
20. 70.8 Ω

12.12   Multiple-choice questions

12M1 (b)
12M2 (d)
12M3 (a)
12M4 (c)
12M5 (a)
12M6 (d)
12M7 (a)
12M8 (b)
12M9 (c)
12M10 (b)
12M11 (c)
12M12 (b)
12M13 (d)
12M14 (a)
12M15 (c)
12M16 (c)
12M17 (b)
12M18 (a)
12M19 (b)
12M20 (d)

Chapter 13

13.13   Exercises

5. 19.05 kV
6. 288.68 V
7. 47.92 A 33 kV

8. 69.28 A
19. 0.32 Ω
20. 20 m

13.14 Multiple-choice questions

13M1 (c)
13M2 (d)
13M3 (c)
13M4 (a)
13M5 (b)
13M6 (b)
13M7 (c)
13M8 (a)
13M9 (a)
13M10 (a)
13M11 (c)
13M12 (d)
13M13 (b)
13M14 (b)
13M15 (b)

Chapter 14

14.8 Multiple-choice questions

14M1 (c)
14M2 (a)
14M3 (a)
14M4 (c)
14M5 (c)
14M6 (d)
14M7 (a)
14M8 (c)
14M9 (b)
14M10 (c)
14M11 (a)
14M12 (d)
14M13 (a)
14M14 (b)
14M15 (d)
14M16 (c)
14M17 (b)
14M18 (d)
14M19 (c)
14M20 (a)

Chapter 15

15.10   Multiple-choice questions

15M1 (a)
15M2 (c)
15M3 (b)
15M4 (d)
15M5 (a)
15M6 (d)
15M7 (b)
15M8 (a)
15M9 (c)
15M10 (c)
15M11 (d)
15M12 (b)
15M13 (c)
15M14 (d)
15M15 (c)

Chapter 16

16.9   Multiple-choice questions

16M1 (c)
16M2 (d)
16M3 (a)
16M4 (d)
16M5 (c)
16M6 (d)
16M7 (b)
16M8 (d)
16M9 (a)
16M10 (c)
16M11 (c)
16M12 (b)
16M13 (a)
16M14 (d)
16M15 (a)

Chapter 17

17.9   Exercises

4. 1.4 V
5. 2 A, 2.1 V

6. 0.2
7. 0.0178
8. 0.025
10. 12 V
11. 3 A, 5.4 V
14. 27 V
16. (a) 2.3 V (b) 3.1 V
17. (b) (i) 5.5 V, 2 A, (ii) 1.1 V, 10 A
18. 36 V
19. 0.4
20. 1.05 V
21. (a) 90 V(b) 110 V
22. 1.4 A, 1.26 V
23. 0.4, 8 V
24. (a) (i) 1.12 A (ii) 8.06 V
    (b) (i) 0.193 A (ii) 1.39 V
    (c) (i) 0.549 A (ii) 3.95 V
25. 80%
26. 66.7%

17.10   Multiple-choice questions

17M1 (c)
17M2 (b)
17M3 (a)
17M4 (d)
17M5 (b)
17M6 (d)
17M7 (a)
17M8 (d)
17M9 (c)
17M10 (c)
17M11 (a)
17M12 (b)
17M13 (d)
17M14 (a)

Chapter 18

18.5   Exercises

1. (a) 816, 544
   (b) 748, 612
   (c) 714, 646
   (d) 693.6, 666.4

2. +6.7%
3. (a) 13.6 mA
   (b) 42.9 mA
   (c) 2.9 mA
   (d) 0.65 mA
   (e) 6.8 mA
4. (a) 330 k ±5%
   (b) 10 k ±10%
   (c) 8.2 M ±20%
   (d) 2.7 ±5%
   (e) 390 ±10%
   (f) 2.2 k ±10%
   (g) 3.9 M ±20%
   (h) 39 k ±2%
   (i) 47 k ±5%
   (j) 5.6 k ±30%

18.6　Multiple-choice questions

18M1 (b)
18M2 (d)
18M3 (a)
18M4 (d)
18M5 (c)
18M6 (a)
18M7 (b)
18M8 (d)
18M9 (a)
18M10 (d)
18M11 (a)
18M12 (c)
18M13 (c)
18M14 (b)
18M15 (d)
18M16 (a)
18M17 (b)
18M18 (c)
18M19 (c)
18M20 (d)
18M21 (b)

Index

absolute permeability 148–9

absolute zero 125

addition 6–7

addressable fire alarm system 281

aesthetic/aesthetics 309

air-cored solenoids 145–8

air thermostat 132

alkaline cells 323, 327–8, 332

alternating current (AC) 165,
178, 199

  advantages of 200–1

  capacitive AC circuit 211–13

  graph of 200

  impedance 214

  inductive AC circuit 210–11

  phasor representation and phase
    difference 206–9

  resistive AC circuit 209

  sinusoidal waveforms 204–6

  values for AC supplies 201–4

aluminium 37, 72, 291–2

ambient temperature 293

ammeters 46, 55

ampere hours (Ah) 43, 337–8

analogue fire alarm system 280–1

annealed copper 74

anode 320

arc 45

arc fault detection devices (AFDD) 45,
261–2

arctic flex 300

atom 31–2

auto fuses 257

BACnet 285

bare conductors 303–4

basket resembles 310

batteries 288, 319–20, 334–7

bipolar junction transistor (BJT) 353–7

bishops 5

block-and-tackle 99

breaking capacity 249–50

BS7671 IET Wiring Regulations 23

Buchholz relay 176

building management systems (BMS)
277, 281, 284–5

building regulations 22–3

bus bar trunking 308–9

butyl rubber 296

cable clips 311

cable conductor manufacture 291

cable insulators 294–6

cable tray 310

cable types 297

  bare conductors 303–4

  data cables 304–5

  fireproof cables 301–3

  flexible cable 299–301

  overhead cables 303

  power cable selection 297

  single-core 297